D0773271

ECOSPEAK

Rhetoric and Environmental Politics in America

M. Jimmie Killingsworth
Jacqueline S. Palmer

Southern Illinois University Press
Carbondale and Edwardsville

Printed in the United States of America
Designed by Duane E. Perkins
Production supervised by Natalia Nadraga

98 97 96 95 5 4 3 2

Library of Congress Cataloging-in-Publication Data

Killingsworth, M. Jimmie.
Ecospeak : rhetoric and environmental politics in America / M. Jimmie Killingsworth and Jacqueline S. Palmer.
p. cm.
Includes bibliographical references and index.
1. Environmental policy—United States. 2. Human ecology—United States. 3. United States—Economic policy—1945- —Environmental aspects. 4. Political culture—United States. 5. Rhetoric—United States. I. Palmer, Jacqueline S. II. Title.
HC110.E5K5 1992
363.7′056′0973—dc20 91-12839
ISBN 0-8093-1750-8 CIP

Portions of chapter 5 appeared in M. Jimmie Killingsworth and Dean Steffens, "Effectiveness in the Environmental Impact Statement: A Study in Public Rhetoric," *Written Communication* (vol. 6), pp. 155–80, copyright © 1989 by Sage Publications, Inc. Reprinted by permission of Sage Publications, Inc.

The paper used in this publication meets the minimum requirements of American National Standard for Information Sciences—Permanence of Paper for Printed Library Materials, ANSI Z39.48-1984. ∞

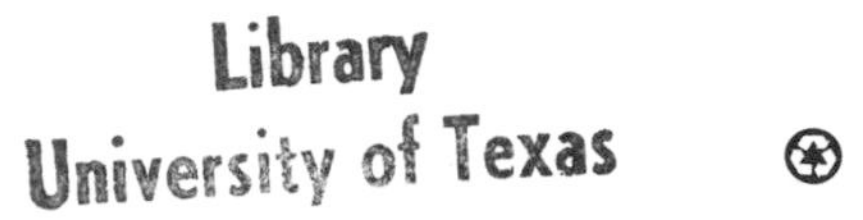

To Miki,
with love and hope

Once upon a time all the world spoke a single language and used the same words. As men journeyed in the east, they came upon a plain in the land of Shinar and settled there. They said to one another, "Come, let us make bricks and bake them hard"; they used bricks for stone and bitumen for mortar. "Come," they said, "let us build ourselves a city and a tower with its top in the heavens, and make a name for ourselves; or we shall be dispersed over all the earth." Then the Lord came down to see the city and tower which mortal men had built, and he said, "Here they are, one people with a single language, and now they have started to do this; henceforward nothing they have a mind to do will be beyond their reach. Come, let us confuse their speech, so that they will not understand what they say to one another." So the Lord dispersed them from there all over the earth, and they left off building the city. That is why it is called Babel, because the Lord made a babble of the language of all the world; from that place the Lord scattered men all over the face of the earth.

—Gen. 11.1–9, New English Bible

Contents

Illustrations

Figures

Table

Acknowledgments

The dilemma over the human relation to the natural world potentially involves every field of academic and practical endeavor. In our effort to understand the relationships among language, thought, and action in environmental politics, we have benefited by the published and unpublished writings of many authors in English studies, education, communication, philosophy, sociology, psychology, politics, and the sciences. We have also been especially fortunate in our conversations with our colleagues and students in these various fields. Those who have left their particular mark on our work include Michael Gilbertson, Dean Steffens, Carolyn Rude, Donald Cunningham, Janet Greathouse, Ron Cleminson, David Lyons, David Hiley, Rex Enoch, John Duvall, Tom Nenon, Mark Gilman, Joan Weatherly, Irma Russell, Sandra Marquardt, Ray Semlitsch, Bill Gutzke, Jim Payne, Steve Klaine, Ron Mumme, Mark Hinman, Beverly Collins, Gary Wein, Ed Stevens, John Haddock, Bill Dwyer, Eddie Seiber, the Memphis Greens, and Tarla Peterson. We thank especially those scientists and activists who shared their writing and took the time to discuss it with us. We cannot claim to have represented their positions as clearly as they themselves could have, for we have seen their words through the lens of our interpretations. Any blame or inadequacy that arises from the views we offer here can only be our own.

We owe a great debt of appreciation to Memphis State University for its support during crucial moments in our research and writing. The English Department chair, William O'Donnell, generously arranged a partial release from teaching during the spring semester of 1989. During the spring of 1990, further release time was provided through a fellowship from the Center for the Humanities under the direction of Professor Kay Easson, who also provided encouragement and guidance along the way.

During the ten-year gestation of the book, we were assisted by many patient and resourceful librarians at New Mexico Institute of Mining and Technology, Texas Tech University, Memphis State University, and Texas A&M University. Deborah Brackstone of Interlibrary Services at Memphis State was especially helpful.

Charles Bazerman and two anonymous reviewers read the entire manuscript in draft and gave valuable advice on how to improve the book. Kenney Withers, Susan H. Wilson, Rebecca Spears Schwartz, and the excellent staff at Southern Illinois University Press also did everything in their power to make it the best book possible. We must ourselves take responsibility for any flaws that remain.

An earlier version of chapter 5 was published as "Effectiveness in the Environmental Impact Statement: A Study in Public Rhetoric" in *Written Communication* (6 [1989]: 155–80). We thank the editors and referees for their help and Sage Publications for permission to reuse the material. The portion of chapter 6 dealing with Earth First! and the Monkey Wrench Gang originally appeared in the article "Realism, Human Action, and Instrumental Discourse," published in the *Journal of Advanced Composition*, 11.2, in 1991.

ECOSPEAK

Rhetoric and Environmental Politics in America

The ferment of contemporary discussions about signs is one evidence of the tensions which beset our culture. Language is of such central importance that it becomes an object of central concern in times of extensive social readjustment.

—Charles Morris,
Writings on the General Theory of Signs (79)

One fact that emerges from a study of the history of rhetoric is that there is usually a resurgence of rhetoric during periods of violent social upheaval.

—Edward P. J. Corbett,
Classical Rhetoric for the Modern Student (32)

[Even] the best-intentioned reformer who uses an impoverished and debased language to recommend renewal, by his adoption of the insidious mode of categorization and the bad philosophy it conceals, strengthens the very power of the established order he is trying to break.

—Max Horkheimer and Theodor Adorno,
Dialectic of Enlightenment (xii, xiv)

The purpose of Newspeak was not only to provide a medium of expression for the world-view and mental habits proper to the devotees of Ingsoc, but to make other modes of thought impossible.

—George Orwell,
1984 (246)

Introduction

Rhetoric and the Environmental Dilemma

Ecospeak and Rhetorical Analysis

Since the middle of the last century, human beings have become increasingly aware of the earth's vulnerability. Examining the results of humankind's technological power has opened a new vein of consciousness, the knowledge that large-scale human action may place the further existence of nature—including human activity itself—in jeopardy. With this new awareness have come calls for "extensive social readjustment" as well as the initial rumblings of "violent social upheaval"—the historical conditions that, according to Charles Morris and Edward Corbett, give rise to an invigorated practice and study of rhetoric.

Classically defined as the production and interpretation of signs and the use of logical, ethical, and emotional appeals in deliberations about public action, rhetoric is both a theory and a practical art. On the one hand, it analyzes and models discourse practices; on the other hand, it seeks to improve these practices. Our purpose in this book is primarily analytical. We want to delineate the patterns of rhetoric typically used in written discourse on environmental politics. We have in mind two audiences, one motivated by scholarly interests, the other guided by personal or political interests. The first audience is composed of students of public rhetoric, to whom we offer a work of *rhetorica utens*, a study of rhetoric in use. We agree with Charles Bazerman that histories of rhetoric have too exclusively focused on *rhetorica docens*, the theory and pedagogy of rhetoric, while ignoring "actual living practice" (Bazerman 15). One of our purposes, then, is to restore the balance in the field by making a practical contribution to the art of rhetorical criticism. The consequences of this pragmatic art, however, may well reach beyond the classroom and the library.

In addressing these consequences, we are brought into contact with our second audience—people engaged in the effort to adjust thought and action to the changing conditions of human life—scientists, government officials, investors, managers, workers, farmers, environmental activists, nature mystics, and anyone else who puts thoughts on paper with the intention of changing the way others think and act in the world. To this audience, we offer a provisional map of recent writers' attempts to reach new stages of consciousness and action through the medium of language. Our sketch of discourse patterns is necessarily rough, because the world and the rhetoric that shapes it are subject these days to strong and unpredictable currents of change. But we feel it is time to make a start. For the audience of activists, we hope our map is useful as a set of hints toward an improved language of public discourse. For scholars in rhetoric, we offer little more than a point of departure for further research.

In these suggestions, we do not mean to claim either that activism and scholarship can be simply and clearly divided or that we ourselves are merely neutral observers. As rhetorical analysts, we are no less users of rhetoric ourselves, and our way of understanding the medium in which we work suggests that, like those whose writings we analyze, we inevitably reveal certain biases in our analyses and evaluations, some of which, owing to the power of ideology, we will never realize ourselves until readers point them out to us. We can say from the start, however, that our sympathies lie primarily with the perspective we will identity as eco-humanism or social ecology, a line of thought best represented in the writings of Aldo Leopold, Rachel Carson, Barry Commoner, Herman Daly, John Cobb, and Lester Brown. The critiques and programs offered by these writers vary greatly, as we will show, but they hold in common the view that technological and bureaucratic solutions to environmental problems will be ineffective—or impossible—unless accompanied or preceded by free and broad access to special knowledges and relevant information as well as by deep psychological and social adjustments. For social ecology, the environmental dilemma is a problem generated by the way people think and act in cultural units. Since human thought and conduct are rarely, if ever, unmediated by language and other kinds of signs, it is understandable—possibly inevitable—that rhetorical scholars enter the environmental discussion through the gate of humanism. Our hope is to remain critical of this perspective and despite our inclination toward it, to weigh it as fairly as possible in the balance with other perspectives.

A Crisis of Western Liberalism: Ethics and Epistemology

An ecologically conscious humanism tends to portray current environmental problems as a crisis of Western liberalism. Briefly put, the dilemma is this: How can the standard of living attained through technological progress in the developed nations be maintained (and extended to developing and undeveloped nations) if the ecological consequences of development are prohibitive? The old liberal concern with material progress has come into conflict with the new liberal concern of environmental protection. This confrontation has led to some serious questions about what constitutes the good life in a progressive culture and about how to reconcile new scientific knowledge with material progress based on information generated by a scientific culture of the recent past. The influential scientist and environmentalist Barry Commoner stated the problem this way in 1976:

> In the last ten years, the United States—the most powerful and technically advanced society in human history—has been confronted by a series of ominous, seemingly intractable crises. First there was the threat to environmental survival; then there was the apparent shortage of energy; and now there is the unexpected decline in the economy. These are usually regarded as separate afflictions, each to be solved in its own terms: environmental degradation by pollution controls; the energy crisis by finding new sources of energy and new ways of conserving it; the economic crisis by manipulating prices, taxes, and interest rates.
>
> But each effort to solve one crisis seems to clash with the solution of the others—pollution control reduces energy supplies; energy conservation costs jobs. Inevitably, proponents of one solution become opponents of the others. Policy stagnates and remedial action is paralyzed, adding to the confusion and gloom that beset the country. (*Poverty of Power* 1)

The environmental dilemma thus creates one of the many situations that the philosopher Alasdair MacIntyre has seen as typical of contemporary Western society's inability to resolve ethical problems. The polis, the ground for agreeable public action, has been divided into separate conflicting interests, each with its own set of values and its own action agenda.[1] The various interests constitute not a unified public, as the political analyst Christopher Bosso has noted, but instead a cacophony of many "publics" (242). As in arguments over abortion, nuclear armament, and population control, participants in the environmental debate appear to work from ethical foundations so widely separated that compromise becomes irrational and conflict endures with no end in sight.

The difficulties of the environmental dilemma are compounded further because the ethical problem issues from a crucial epistemological problem—humankind's "alienation from nature." The relation of person to person or person to community—the province of ethics—overlaps into and comes to depend upon the relation of person to world. In all ethical problems, we must consider the rules for community formation, but in environmental disputes, we must additionally understand how the disputants construct their views of the natural or nonhuman worlds. One group will view nature as a warehouse of resources for human use, while an opposing group will view human beings as an untidy disturbance of natural history, a glitch in the earth's otherwise efficient ecosystem. Between such extremes, there are any number of conventional or idiosyncratic constructions of the person-planet relation. Writers on environmental issues must therefore hope to influence not only their audience's ethical attitudes but also the way the reader regards the entire community of nature.

Ultimately, any analysis that follows the trend of modern Western individualism in dividing consciousness into self versus world or self versus community will prove psychologically inadequate. In *The Natural Alien*, Neil Evernden uses Heidegger's phenomenology to suggest the ultimate flaw in the liberal, or individualistic, understanding of the relationship of the self and its environment. Despite the analytical tendency of Western thought since Plato, experience, if carefully attended to, suggests that we are not divided against our world but rather that human consciousness *contains* the world. As Heidegger argues, the experience of being is always one of *being-there (Dasein)*. For Evernden, a person "does not really experience the boundary of the self as the epidermis of the body, but rather as a gradient of involvement in the world" (64). To divide self from world is to divide consciousness against itself. In this sense, the environmental dilemma is as much an internal, psychological problem, a problem of self-alienation, as it is an external problem involving the self's control over its external resources. In *For the Common Good*, Herman Daly and John Cobb argue along the same lines that "in the real world the self-contained individual does not exist" (161). The primary target of their critique is the human experience as constructed by modern economics, the image of *homo economicus*, an independent, pleasure-seeking consumer, a person-in-body. Against this image, Daly and Cobb posit an image of the human self as a social being, a person-in-community: "We come into being in and through relationships and have no identity apart from them. Our dependence on others is not simply for goods and services. How we think and feel, what we want and dislike, our aspirations and fears—in short, who we are—all come into being socially" (161). Contrary to the ideology of the "self-made" woman or man, then, the self is socially constructed, never entirely independent of its social context (Daly and Cobb) or its natural context (Evernden).

The social and psychological dimensions of this crisis in liberal consciousness stand out clearly in Samuel Hays' historical study of public environmental awareness from 1955 to 1985. "The search for environmental quality," writes Hays, "was an integral part of the rising standard of living. Environmental values were based not on one's role as a producer of goods and services but on consumption, the quality of home and leisure. Such environmental concerns were not prevalent at earlier times. But after World War II, rising levels of living led more people to desire qualitative experiences as well as material goods in their lives" (*Beauty* 34–35). According to Hays' analysis, environmental awareness is, like increased demands for improved housing and public health, a form of consumer activism, distinctively a public movement, an issue connected with the distribution of social wealth. Viewed psychologically, there is some irony in this interpretation. The very social system that provides a higher standard of living—

the technologically supported, consumer-oriented free market—also produces the pollution that reduces the quality of life. If we follow Hays (as well as Herman Daly, Lester Brown, and other ecological economists) in categorizing environmental quality as yet another "amenity" of modern life, to be counted with other amenities like labor-saving devices, fast automobiles, and large houses, then we can say that the techno-capitalist system of production is able to provide certain amenities only at the expense of others. Moreover, as Hays notes (following other observers, notably Barry Commoner), environmental amenities (like all amenities) are unequally distributed: "The more affluent . . . live on the fringes of the city away from the noise, polluted air, and congestion of the more densely settled parts," while "the urban poor . . . bear the burden of air pollution" and other unpleasant conditions associated with proximity to the point of production (*Beauty* 266–67).

As more and more people enter the lists of the affluent, pressure on better living spaces and recreation spots free of technological taint increases. And as environmental degradation becomes more and more pervasive—involving such global concerns as acid rain, the greenhouse effect, and the pollution of the oceans—the general public is mobilized to the cause of environmentalism. At this point, the public must face the dilemma: Does the system that feathers the nest create conditions that ultimately make life unpleasant or unbearable? Can the system be adjusted to compensate for the side effects of progress, or must the system be thoroughly overhauled, if not replaced? What are the limits to the liberal faith in technology?

A Crisis of Western Liberalism: Discourse

As much as the environmental dilemma is a problem of ethics and epistemology, it is also a problem of discourse. Various proposals to resolve the crisis are put forth by different social groups with different sources and kinds of information, groups with divergent goals, methods, values, and epistemologies. All groups have a particular perspective and use a specialized language developed specifically to describe and stimulate the practices characteristic of their particular outlook on the world. This specialization of language is clearest in scientific disciplines, where subtle theoretical and methodological distinctions are embedded in working vocabularies. But as contemporary discourse theory teaches, all groups—whether formal or informal—have their specialized languages. In the course of a single day, anyone might use several different kinds of language to indicate membership or to influence actions in several distinct "discourse communities."

In recent years, scholars have begun to concentrate on the internal rhetoric of discourse communities (see Latour, for example). This work substantiates and extends the study of general rhetoric, which has traditionally been concerned with the specialized languages of discourse communities only as they enter into the shared space of public debate and as they dissolve into the loosely agreed upon language of the marketplace and the mass media (Barilli). On this common ground, where the turf of public life is divided and territorial conflicts are enacted, writers and speakers create *appeals*, the rhetorical techniques used to construct relationships between different groups for the purpose of cooperative social action. The chief end of this general, public rhetoric is *identification*, as Kenneth Burke has shown in *A Rhetoric of Motives*. For the rhetorical analyst, the intractability of social problems like the environmental dilemma is due to the inability of concerned discourse communities to form adequate identifications through effective appeals. One such inadequacy, for example, is the failure of environmentalist groups with broad social programs to capitalize on the growing public concern about the environment and thereby to set their programs permanently in motion. What is true for the government-sponsored social engineers that grew up in the conservation movement at the beginning of this century (see Hays, *Conservation and the Gospel of Efficiency*) is true also for newer groups like the deep ecologists, wilderness preservationists, eco-anarchists, and green politicians: They have been unable to create strong communicative links with the mass public, links that would support a strong power base for reformative actions.

Following Burke, we can think of this problem in grammatical and narrative terms. Any action may be stated by an active voice sentence, the kernel of a group's identifying story: *I (or we) do this*. In the case of intractable problems, the subject position of one group ("we") cannot be filled with members of another group ("you" or "they").

Rhetorical appeals propose enlargements of the *we* category or mergers of two or more categories, with the ultimate goal being the identification of the "global" public with the "local" discourse community. Failure of such appeals ends in divisions that harden with time and with the repetition of rhetorical situations whose narrative outlines contain similar plots and characters—the confrontation of environmental activists with land developers, for example. Familiarity breeds oversimplification, stereotyping, and pigeonholing—especially in the mass media, where a telegraphic style mingles with the need to cultivate in the audience a quick recognition of issues and public figures.

Now we come to the region of *ecospeak*, where public divisions are petrified, conflicts are prolonged, and solutions are deferred by a failure to criticize deeply the terms and conditions of the environmental dilemma. Ecospeak has emerged as a makeshift discourse for defining novel positions in public debate. Advocates of environmental protection, for example, have come to be called, innocently enough, *environmentalists*. This word and the habits of thought associated with it have become part of a vocabulary that fills not only our newsmagazines but also our minds. We may add the category "environmentalism" to the mental scorecard we use for politicians running for office. Or we may worry about whether we can be "good environmentalists" and still support the development of local technologies that provide jobs for our neighbors and appear to make our lives easier. Or we can consign a local problem with water pollution to the pigeonhole "environmental problems," and if we are inclined to other areas of specialization or interest, we may then leave that problem or "issue" to the "environmentalists."

In one sense, the environmental dilemma only reflects the current historical stage of ecological consciousness; the environmentalist ethos is only now in the process of formation so that the conflicts we feel are part of the inevitable process by which a political stance evolves. In another sense, however, we may worry that as this new ethos solidifies, its characteristic terminology and working rhetoric will cause oppositions to close into an irresolvable set of conflicts, much as, in the debate on abortion, the pro-life ethos has established—once and for all, it seems—argumentative positions diametrically opposed to the pro-choice ethos, and vice versa. Like *Newspeak*, the austere vocabulary of mind control in Orwell's politicolinguistic fable *1984*, ecospeak becomes a form of language and a way of framing arguments that stops thinking and inhibits social cooperation rather than extending thinking and promoting cooperation through communication. A political writer interested in building constituencies through identification thus takes a profound risk in using ready-made terminology and engaging in old quarrels because, as the critical theorists Horkheimer and Adorno have warned, "[even] the best-intentioned reformer who uses an impoverished and debased language to recommend renewal, by his [or her] adoption of the insidious mode of categorization and the bad philosophy it conceals, strengthens the very power of the established order he [or she] is trying to break" (xiv).

Ecospeak has produced its own rhetorical analysis of environmental politics, which emerges in the mass media and in ordinary conversation as an oversimplified dichotomy. On one side are the environmentalists, who seek long-term protection of endangered environments regardless of short-term economic costs. On the other side are the developmentalists, who seek short-term economic gain regardless of the long-term environmental costs. This analysis oversimplifies the dilemma by projecting the psychological dilemma—the realization that our system produces both economic prosperity and environmental pollution—onto a social background, dividing two stages of liberal consciousness against one another in a kind of allegory of good guys and bad guys, demanding of the observer a value judgment about the goodness or badness of each side. This representation has its roots in the first conflicts over land use at the end of the last century, when Americans found themselves overtaking the last bits of the continental frontier. As the historian Stephen Fox has suggested,

> The campaign for Yosemite set a pattern to be repeated many times in future public quarrels over the environment. At stake: a piece of the natural world. On one side: its defenders, spearheaded by amateurs with no economic stake in the outcome, who took time from other jobs to volunteer time and money for the good fight. On the other side: the enemy, usually joining the struggle *because of* their jobs, with a direct economic or professional interest in the matter and (therefore) selfish motives. Politics seldom lends itself to such

> simple morality plays. But environmental issues have usually come down to a stark alignment of white hats and black hats. (Fox 104)

This "stark alignment" is rarely a matter of historical necessity, however, but rather a device of discourse used by one side or the other (often both) to mobilize forces against a palpable villain. A more complex view of the rhetorical situation is risky for either side, because it could result in what Jesse Jackson likes to call "the paralysis of analysis."

But oversimplification itself, for all of its focusing power, offers no warranty against political paralysis. The wilderness ethic and its dichotomous rhetoric have not produced impressive victories in the national and international arena of environmental politics; and even at a local level, the history of wilderness protection is a story of strong rhetoric followed by embarrassing and often demoralizing compromises. The tendency to divide into two narrowly defined parties often leaves a huge population untouched or confused by the debate and the ensuing actions. Moreover, as the division is reported in the news media—where the environmentalist/developmentalist dichotomy has proved immensely attractive and durable, presumably because it provides busy reporters with a ready-made stock of plots and characters—it tends to conceal other sources of solidarity and conflict, which if closely examined, could provide hints toward the kind of social reorganization needed to cut through the environmental dilemma.

Further rhetorical analysis breaks the hold of ecospeak by identifying various discourses on the environment before they are galvanized by dichotomous political rhetoric. It does so too by studying the transformations of these discourses as they enter the public realm by way of a local discourse community (whether a professional ghetto like "the scientific community" or an actual region defined by geographic and democratic features). At the very least, such analysis can reveal possible identifications and real conflicts passed over by an ever-too-glib retreat into ecospeak.

Mapping the Discourses

As a first step beyond the simple dichotomy of environmentalist versus developmentalist, we will analyze in the following chapters a number of works by various authors representing several distinct ethical and epistemological perspectives on environmental issues. For the remainder of this introduction, we will develop a simple framework to use as a point of departure in tracing the relationships of the different discourse communities that contribute to each author's work. This model is offered tentatively. Though it challenges the oversimplifications of ecospeak, it nevertheless remains itself oversimplified and schematic, drawing rather too heavily on many of the categories of ecospeak, but nevertheless arranging them continuously, thus suggesting relationships that cut through the absolute divisions of ecospeak. In making and using such models, it is helpful to remember a comment of the linguist Kenneth Pike: "Only if a theory is simpler than that reality which it is in part reflecting is it useful" (6). Later chapters will reveal both the usefulness and the ultimate inadequacy of this simple map. For now, it will at least serve as a means of introducing the major players in the public debate on the ecological crisis in America.

A *Continuum of Perspectives on Nature*

Figure 1 plots the major perspectives along a continuum whose poles designate three human attitudes toward the natural world.[2]

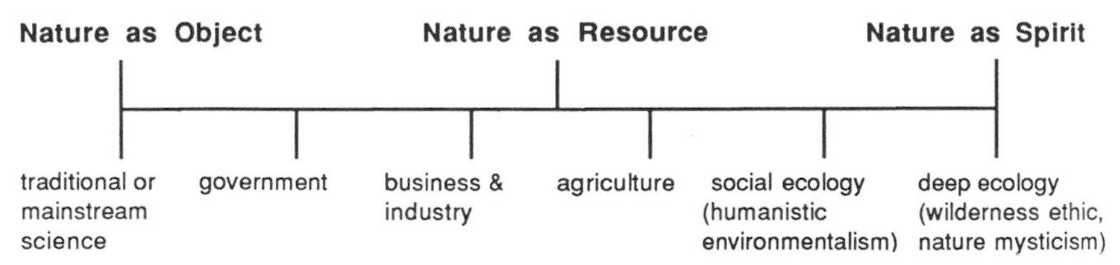

Figure 1. Continuum of Perspectives on the Environment

The continuum offers an alternative polarity to that suggested in ecospeak. Instead of placing developmentalism (nature as resource) as the absolute opposite of environmentalism or deep ecology (nature as spirit, or *presence*), the map extends in two directions away from the center, placing at one extreme the perspective of traditional science, which, due to its concern with objectivism and its rejection of anthropocentric thinking, would depart from the view of nature as a resource.

We are thinking here of experimental science as it has developed since the seventeenth century, with its fabled detachment from all natural objects (including human beings). This view of science has been encapsulated and rigidified in government and industry in the form of "scientific management." At the opposite pole is "deep ecology," which departs from both objectivity and anthropocentricity, asserting instead a mythic involvement with nature, an identity in which the spirit of creation wraps the human and the nonhuman in an indissolvable unity with definite ethical consequences. From the anthropocentric perspective at the center of the continuum, which holds that nature is a bounty of resources for human use and enjoyment, both the scientific and the deep ecological outlooks could prove threatening. And indeed history has joined the two ends against the middle, as we will show. But first let us consider a few complications of the scheme.

Any person has almost certainly experienced all of the various attitudes toward nature listed above the continuum, even if he or she has but a limited experience with the different, more specific outlooks listed underneath the continuum. Despite this common human breadth, however, at any given moment, one attitude may dominate the others. A botanical scientist, for example, may behold a rare orchid with the eyes of a bona fide objectivist in the laboratory, or with the pride of a gift-giver or the pleasure of the recipient of the gift, or finally, with the bliss of a nature mystic on a jungle expedition. Ecospeak and related forms of political rhetoric seek to achieve a measure of control over an audience or an opponent in debate by categorizing the opponent into a single role assigned on the basis of a dominant attitude. Moreover, debators may seek shelter from attack by claiming for themselves a breadth of attitudinal experience denied them by their opponents. A businessman interviewed in a recent article in *Business Month* (Haines), for example, may be forgiven for missing the point entirely when having been marked as an enemy of green politics, he protests that in fact he is a great lover of nature. It is, of course, his business that causes the problem, not his personal attitude.

Considering the continuum from a pragmatic orientation would help to prevent this kind of mistake. Each of the basic attitudes could be aligned with a general field of action. First, the view of nature as object is most useful in scientific research because it allows for quantitative analysis. If things are to be counted, then there must be things, even if the very existence of these entities is placed in question by scientific theory. In this practical view, action (as defined by ethics) tends to be reduced to motion (as defined by physics), and all relationships are reduced to correlation, cause, and effect. If such reduction is impossible (as, for example, in matters of aesthetics and ethics)—if number values cannot be assigned—then science must yield to softer forms of analysis. Second, the view of nature as resource is most useful in the transactions of the marketplace, where once again counting is paramount and as a consequence, natural entities are reduced to objects that can be priced as commodities. The concept of ethical action is retained but is applied exclusively to human behavior, with the rest of the world reduced to meaningless motion. Third, the view of nature as spirit places human beings on a par with the rest of nature, extending ethical action to all beings of the earth. The characteristic actions prompted by this attitude—beyond prayer, meditation, and bearing witness—involve an active resistance to the other perspectives that violate that identity of human beings and nature.

Along the continuum, political groups tend, as it were, to form outposts that are from time to time uprooted and restaked according to the formations and transformations of new alliances. The best example is government, which moves this way and that according to the orientation of a particular congress or a particular administration. Pragmatically considered, however, government can never be absolutely identical either with science or with business. Governmental action is devoted primarily neither to understanding nor to making money. Above all, it is devoted to control and to the self-perpetuation of a social system. Scientific understanding is useful only if it empowers government to control and perpetuate the system. Likewise (all cynicism aside), the government raises money not for the pleasure or glory of its employees, but for the effective maintenance of the system. Thus for government, nature may become *either* an object of study *or* a resource to be managed, but neither is the ultimate aim of government institutions.

A typical analysis of the capitalist hegemony, or power base, indicates a harmonizing of science, government, and business made possible by a common tendency to use technology as a means of molding the world into productive systems, to produce knowledge in science, military strength and information ("intelligence") in government, and money in business. With the environmental crisis, however, this complex of power relations has loosened and threatens to unravel, primarily because the strength of the environmentalist appeal to science has coincided with a theoretical trend among scientists toward ecological holism, which is rather surprisingly sympathetic with a view of nature as spirit. Under the pressure of developments over the last hundred years, the Enlightenment continuum that placed science and nature mysticism at opposite poles now bends into a horseshoe, which brings science and deep ecology into a closer relationship. Figure 2 is an attempt to render these developments schematically. The direction of rhetorical appeals is shown by arrows in the figure.

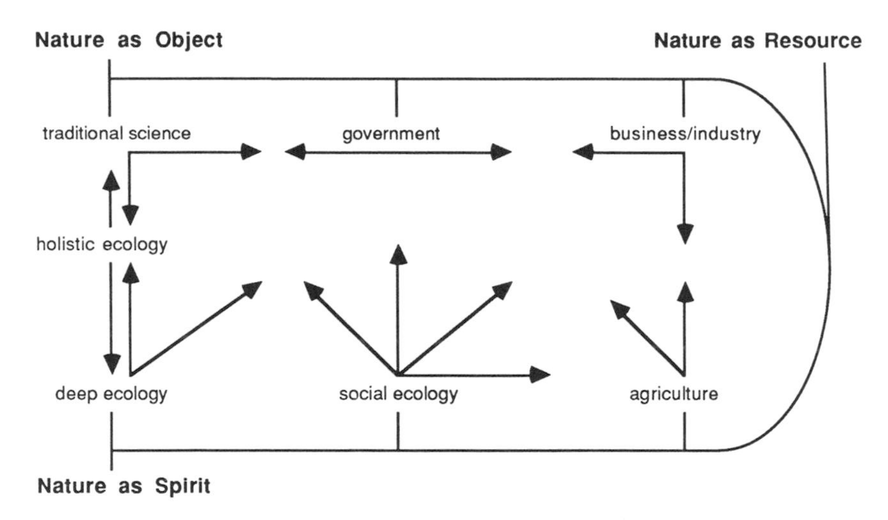

Figure 2. Horseshoe Configuration of Perspectives

This plotting of the perspectives visually suggests four concepts important to our analysis of the rhetoric of environmental politics—hegemony, opposition, tension, and direction of appeal.

Hegemony. On the upper axis appear the perspectives that have enjoyed the greatest success in winning public support and power—science, government, and industry. Their greatest glory came in alliance with one another, potently symbolized in the Manhattan Project and the continued development of the scientific-military-industrial complex after World War II. The hegemony of this triumvirate is threatened by the discourses that appear on the lower axis, all of which lack sufficient political strength to put forth their programs except as amendments to the programs of the upper axis groups. Historically, the dominant hegemony of science, government, and business has not taken environmentalism very seriously, as Hays has shown: "Among institutional leaders in government, in business, and in the scientific and technical professions environmental values never took deep root. Those leaders were far more preoccupied with more traditional forms of economic growth. . . . By 1980, two-thirds of the public expressed positive environmental concerns, whereas only one-third of the nation's leaders did" (*Beauty* 60).

We may now, however, be witnessing an attitudinal shift and a corresponding power shift that would cause the continuum to "roll," leaving a new alliance of deep ecology, science, and government—the environmentalist alliance—on the upper axis. Such a shift depends largely upon rhetoric, the building of "discursive links," the opening of new "subject positions"; for, as the social theorists Ernesto Laclau and Chantal Mouffe have recently argued, "hegemony supposes the construction of the very identity of social agents, and not just a rationalist coincidence of 'interests' among preconstituted agents" (58).[3] For science to form a hegemonic link with deep ecology or social ecology, for example, it would have to be a transformed science, not the positivistic science that formed the model for scientific management and that provided the impetus for large-scale technological development.

Opposition. In the figure, opposition is shown by distance, both vertical and horizontal. Vertically, the relatively powerful, high-status perspectives—as traditionally formulated in science, government, and business/industry—stand in opposition to the power-seeking or reduced-status perspectives situated along the lower axis. Horizontally, the loose confederation of science and deep ecology—which with strong appeals to government, forms the emerging hegemony popularly known as environmentalism—stands in opposition to the human arts and services clustered around the nature-as-resource pole, the rallying point of the developmentalist resistance as it has developed from Gifford Pinchot to Ronald Reagan (see Hays, *Conservation*; and *Beauty*). The distance between science and industry remains the same as it was in the earlier historical period (indicated in figure 1), while the distance between science and the reformist and radical ecologies has narrowed due to the intervention of holistic thinking and the forcefulness with which human influence upon natural systems has asserted itself. Natural science, as we will show, has never been closer to merging with social science than in the field of general ecology (see Toulmin). Though figure 2 shows a mutuality of interest, the scientific establishment tends to be characteristically reticent in forming this alliance and sharing its authority with deep ecology or social ecology. Nascent theories that could close the gap and turn the horseshoe into a circle, such as the Gaia hypothesis (the concept of the earth as literally a self-regulating organism) and holistic versions of general systems theory, are still consigned to the margins of the accepted canon of knowledge. (Indeed, the author of the Gaia hypothesis, the self-proclaimed "radical scientist" James Lovelock, has charged that things have changed since Galileo struggled with the powers of religious inquisition: Now "the scientific establishment . . . makes itself esoteric and is the scourge of heresy" [Lovelock xiv]).

The horizontal opposition (with the poles more or less equivalent to the environmentalist/developmentalist conflict) is mediated by government, now placed in the central position, the object of everyone's appeal. Both sides would likely agree that government, especially the federal government, has become (for better or worse) the key player in the environmental dispute, the institution with the power to regulate ecological research, environmental action, and development of resources.

Tension. In this figure, proximity—either vertical or horizontal—indicates not only possible links or substitutions, but also strong tension, as our brief analysis of the relationship between science and deep ecology suggests. Social ecology shares with deep ecology a critique of capitalism and a commitment to environmental protection, but it is driven by a deep-felt commitment to human culture and sustainable technology rather than by a mystic communion with nature. Social ecology finds deep ecology's mystifications of the human-nature relation ideologically suspect, and deep ecology mistrusts social ecology's commitment to a continued, though modified, version of the human technological project.

Tension radiates from relationships based on competition or dependence. Both science and business depend upon government intervention, but the two fields greatly resent their own dependence and the intervention itself. They strive to deny or slip free of government's deep influence upon their characteristic activities. In a similar point of tension, agriculture both depends upon and supplies vital business for the banking industry, but it does so in a mood of profound mistrust—a mistrust that dates back centuries, to a time when business was confined to the city, farming to the country (the cultural condition that explains why agriculture joins the two ecological perspectives on the lower axis of the horseshoe, the earth-conscious axis, which stands in cultural opposition to the upper axis with its city-centered institutions—see Bookchin, *Remaking Society*). These tensions charge the atmosphere of American environmental politics and partly define the nature of typical rhetorical appeals.

Direction of appeals. The arrows on the figure indicate the general "direction" of typical rhetorical appeals, the efforts to overcome oppositions and divisions either by forming new solidarities, by reinforcing old ones, or by revealing distances or likenesses in order to transform attitudinal conflicts into political action. The appeals are indexes of power or status; the recipient of many appeals tends to be the strongest institution of a society, the perspective that captures the imagination of the general public, or the power that exerts the greatest influence. It is not surprising that in America, a nation founded on Enlightenment principles, science has drawn a wide range of appeals. It seems somewhat odd, however, that appeals for government intervention have become so common and so widespread in writings on the environment. Americans are said to value their independence and local control of decision making. Nevertheless, in this matter, our findings coincide with those of Samuel Hays; though much environmentalist rhetoric affirms a "decentralist view," environmental activists ultimately seem unable to free themselves from dependence upon a centralized federal authority (*Beauty* 247).

The Plan of the Book

Our task from this point on will be to demonstrate how, through the use of rhetorical appeals, the various perspectives attract and repel one another and how identifications are formed through the merging of communities and their characteristic styles and genres.

Chapter 1, "Varieties of Environmentalism: A Genealogy," will present an analysis of polemical rhetoric in the political action groups that have emerged since the late nineteenth century. Our thesis is that, as the environmental crisis has intensified, the identities of special interest groups and their corresponding rhetorics have tended to undergo strong dislocations, with new groups emerging from old ones to represent each new stage of a deepening political consciousness. The older groups do not, however, just die off; they remain in existence as points of tension for the new groups. Conservationism grew up in rivalry with its younger sibling, reform environmentalism, which sought a biocentric rather than an anthropocentric outlook on change. Eventually this more robust and far-reaching perspective spawned deep ecology and social ecology, which differ from reform environmentalism primarily in demanding that political reform be accompanied by widespread attitudinal and behavioral change as well as extensive social reorganization. With this development comes a corresponding shift in the history of environmental rhetoric. As the movement grows away from local and small-group politics toward global interests, it tends to abandon agonistic, single-issue rhetoric, adopting instead a social-epistemological, systematic presentation of action agendas, the rhetorical precedents for which are found primarily in the realms of religion and international politics.

Chapter 2, "The Rhetoric of Scientific Activism," extends the historical narrative of chapter 1 through a treatment of the three major figures responsible for building strong alliances between environmental activism and scientific learning from 1950 to 1980—Aldo Leopold,

Rachel Carson, and Barry Commoner. These writers brought together the new political consciousness with the paradigm of scientific holism that had begun to emerge in the 1950s. We will argue that the effort to unify scientific ecology and social ecology, which correlated historically with the spread of environmental consciousness, gave new status and power to environmentalist arguments, but left the exact nature of social change uncertain. An index of the influence of scientific activism within the scientific community itself is the extent to which science textbooks used in high schools and colleges since the 1970s have taken up social themes in chapters on ecology. The chief problem with the efforts to fuse scientific ecology with social or deep ecology is that the new forms of thinking are couched in the old rhetoric of earlier periods. Scientific activism has provided a strong critique of current social practice without providing an adequate (instrumental) discourse of social adjustment. From the perspective of rhetorical analysis, the articulation of new positions thus tends to reinforce the paralyzing dichotomies of ecospeak, above all pitting environmentalism against developmentalism and statism against capitalism.

Chapter 3, "Scientific Ecology and the Rhetoric of Distance," attempts to show that despite the appeals to science offered by various political advocates and despite the public influence of scientific activism, the formal discourse of mainstream science preserves its "contemplative" distance through careful restrictions on language and genre and thereby remains ironical on political questions, creating a kind of apolitical politics out of its vaunted neutrality. We trace the traditional rhetoric and approach to scientific activity in the work of several ecologists working in a typical state university.

Chapter 4, "Transformations of Scientific Discourse in the News Media," compares the treatment of scientific findings in the mass media with the way such findings are treated in mainstream science. Whereas science must theorize to survive, science news must sell to survive and thus depends heavily upon the conventions of rhetorical and poetic discourse as shaped by general human interest. To demonstrate some ways scientifically generated information is transformed to meet the needs of the different social groups engaged in environmental politics, we analyze how magazines with distinct target readerships cover stories related to the problem of global warming.

Chapter 5, "The Environmental Impact Statement and the Rhetoric of Democracy," analyzes the genre of writing used by government in an effort to blend scientific expertise and democratic participation in environmental policy decisions. Ironically, the EIS process has drawn fire from both scientific experts and various unrepresented groups. We trace the dissatisfaction to an incomplete resolution of the instrumental, rhetorical, and scientific aims of the EISs in our case study.

Chapter 6, "Rhetoric and Action in Ecotopian Discourse," explores the social and psychological meaning of actions undertaken by groups like Earth First! and Greenpeace, actions used as a means of protest or as community-forming and protopolitical symbols. We also treat a pair of environmentalist novels—Edward Abbey's *Monkey Wrench Gang* and Ernest Callenbach's *Ecotopia*—as fictional models for alternative actions, representations of possible futures for environmental politics. In direct and in symbolic action, utopian discourse creates a space for the imaginative interplay of poetic and rhetorical activities. It posits social links and subject positions beyond the normal conception of the practical and the probable, beyond the petrified mythos of ecospeak. The ecotopians therefore effectively challenge conventional thinking even when they fail to win consent for particular reform programs.

Chapter 7, "Ecological Economics and the Rhetoric of Sustainability," provides first a brief introduction to attempts over the last twenty years to adjust formal economics to the requirements of the new ecological awareness. We focus in particular upon Herman Daly's use of humanist rhetoric to argue that the analysis of ends should be combined with analysis of means in science, social science, and ethics. In the same humanist vein, we analyze Lester Brown's effort to sustain a continuous and pluralist narrative about the physical and moral state of the world.

With this image of an emerging discourse that brings a plurality of knowledges cogently and coherently to bear on a problem afflicting the world public, we conclude on a hopeful, if cautious, note. Even if we find ourselves in a Babel of discourse communities, each with its own characteristic language, epistemological outlook, and agenda for action, there remains in rhetorical inquiry a need, a mission, and a hope for a generally accessible narrative, the story of how human action reconciles conflicting demands in the search for the good life. Even while stressing caution about claims over the accessibility of information, rhetorical criticism urges continued development of the story of human cooperation.

Why, it may be asked, should such a diversity of folk be bracketed in a single category? Because each, in his own way, is a hunter. And why does each call himself a conservationist? Because the wild things he hunts for have eluded his grasp, and he hopes by some necromancy of laws, appropriations, regional plans, reorganization of departments, or other form of mass-wishing to make them stay put.

—Aldo Leopold,
Sand County Almanac (282)

The consequent industrialization of man can be inverted only if the convivial function of language is recuperated, but with a new level of consciousness.

—Ivan Illich,
Tools for Conviviality (96–98).

With a word he shall slay the wicked.

—Isaiah 11.4

1
Varieties of Environmentalism: A Genealogy

The Environmentalist as the Other

Kenneth Burke's chief contribution to rhetorical theory was the concept of *identification* as the means by which a speaker or writer puts forth an image or character—what the ancient rhetoricians called *ethos*—and invites the audience to participate in a consubstantial relationship with that image. The analytical question to which this approach leads us in rhetorical criticism is, "Who is the person, or *who does the person claim to represent*, who puts forth arguments about the environment?" A variant question, one with great significance in the classical rhetorical situation, in which two opponents debate a question for an audience of judges, is this: "Who is the person described in the discourse as the opponent, the challenger, the other?" For usually, identity is revealed through efforts to build communities in the face of conflict. In this chapter, we will apply the question to our topic by asking, "Who is called an 'environmentalist,' by whom, and to what advantage?"

Our analysis reveals a tendency of political identities to develop in response to the experience of actual physical crises. Once these political positions have evolved, they tend to remain in place even as new ones come to the fore, resulting in a proliferation of perspectives. At the end of the last century, when Americans first realized the limits of the vast continental frontier, the conservation movement emerged. In its dominant version, it stressed the efficient use of natural resources—the gospel of efficiency, as Samuel Hays has called it. Under the leadership of men like Theodore Roosevelt and Gifford Pinchot, this approach to the natural world was embodied in government institutions like the Department of the Interior and in new academic disciplines like forestry and wildlife management. Conservationism had a strong appeal, for its rhetoric was typical of the progressive period, a utilitarian discourse boasting of the powers of professional scientific management and disparaging of the amateur, transcendentalist strains of the emerging wilderness movement led by John Muir and his Sierra Club. The roots of ecospeak lie in rhetorical differences between Muir and Pinchot, reflective of their different personalities and their different approaches to the problems of resource use. The victory in their struggle for power went early on to Pinchot. As we will show in chapter 5, his concept of applying scientific management and executive control based on expert analysis prevails to our day in the practices of government agencies like the Forest Service and the Bureau of Land Management.

The fire of the wilderness ethic—the claim that nature has rights of its own, that it is good in and of itself, independent of its resource value—was nurtured by small groups of enthusiasts on the margins of American life. After World War II, the transcendentalist language of wilderness preservation began to have a stronger appeal for the American public, as the conservation movement gave way to reform environmentalism. Whereas the conservation movement had been "an effort on the part of leaders in science, technology, and government to bring about more efficient development of physical resources," as Hays suggests, the newer environmental movement "was far more wide-spread and popular, involving public values that stressed the quality of human experience and hence of the human environment." Thus "conservation was an aspect of the history of production that stressed efficiency, whereas environment was a part of the history of consumption that stressed new aspects of the American standard of living" (*Beauty* 13). The high-minded transcendentalism of Thoreau, Muir, and their followers therefore came to serve the interests of the growing environmental consciousness of consumers who sought healthy environments in which to live and pristine environments into which they could escape from the routine grind of urban, industrial life. The rhetoric of reform environmentalism was also based on appeals to government, thus reawakening the Muir-Pinchot battles of the turn of the century over who would control the direction of the environmental protection effort—scientifically trained experts or concerned amateurs.

Reviving the old rhetoric has led, however, to the encrustation of ecospeak which we outlined in the introduction. The new wine has been put into old wineskins, with ill-fated results. Moreover, activists' attempts to bring the public to a full appreciaton of the rights-of-nature ideology or to redirect public environmental consciousness into new discourses or to link existing ideologies—revised socialisms or leftwing anarchism—to ecological concerns have had limited success. For as Hays rightly argues, "Among only a few environmentalists did action come from general ideologies, larger managerial motives, or more detached analyses of national and international problems. Most became aroused through personal circumstances and immediate needs, shared with others through common experience" (*Beauty* 65).

The political effectiveness of environmentalist rhetoric has thus depended upon a discourse's ability to create valences, open links that attract individuals among the general public by realistically mirroring the experience of daily life without seriously challenging either the basic institutions and ideologies of American life or the values of consumer mentality. The working vocabularies of the old movements remain available; new discourses arise to challenge the old; new communities enter the debate as the effects of pollution are more widely experienced. For better and worse, a general pragmatism and an open, democratic system prevail through flare-ups of capitalist assertion, government intervention, and ideological experimentation. The general public, the force that tips the political balance, assesses the arguments, tests them against its diverse experiences of contemporary existence, and becomes the ultimate "decision-maker" in the political process as we know it. The drama of rhetorical appeal in American environmental politics is thus a contest to win the favor of this mass public by creating language that stimulates first consent and then identification. Which version of environmentalism, we may ask, has this power?

This, however, is not the question ordinarily posed by the news media. Its reliance on the categories of ecospeak leads to a simpler version of the question: Will the American people accept environmentalism? The movement is portrayed as static, settled, essentially identical to the wilderness preservation movement as it began over a century ago. We will argue that this tendency is far from being a neutral mistake, but is rather the result of developmentalist ideology, the work of the opposition, which despite the growth of public support for the environmentalist causes, remains dominant, especially in regional pockets where newspapers are tied to chambers of commerce and other purveyors of economic development in the traditional, industrialist mode.

Ecospeak in the News: A Case

In this traditional mode, the news media have cast environmentalists as the spoilers of economic development in a rhetoric that covertly (or unconsciously) upholds the values of industrialization and the creation of new jobs in a community, regardless of the nature and source of those jobs. People need to work; they need money; they need to satisfy their most immediate needs. That much is clear. The values of environmentalists who oppose job-creating industry, on the other hand, appear in news accounts as mysteriously unfocused. Like all images of the *other*, the resulting portrait of the environmentalist comes to be chiefly characterized by the negative. The environmentalist is anti-development, antiprogress, against the building of this road or that power plant.

Consider, for example, a feature story printed in the 18 June 1989 issue of the *Memphis Commercial-Appeal* (a publication whose very name suggests an inherent rhetorical bias toward developmentalism, an appeal to commerce). The story reports on the industrialization of Calvert City, Kentucky, once a village of fewer than 400 people, but transformed since the 1940s into a center for heavy industry. B. F. Goodrich, GAF Chemical, a steel plant, and an electroplating operation, among other manufacturers, have set up shop in Calvert City. The companies were drawn to the area by an aggressive recruitment campaign that hinted at cheap and plentiful labor and promised cheap power from the Kentucky Dam that had been recently constructed by the Tennessee Valley Authority. Everything was fine until some people began to suggest that the heavy air around the town may be causing the cancer rates to be unusually high. Now the town, as the journalists like to say, is torn by conflict.

The news story, "Pollution a Price as Town Prospers," is outwardly an objective and balanced report on the situation. Every effort is made to give a fair rendering of the town's story from the perspectives of both environmentalists and those who favor industrialization. The objectivism fails, however, for two reasons. First, the environmentalists are named as such and are thus labeled as the others, the newcomers, the spoilers; while those who favor industrialization—the developmentalists—go unlabeled in the article and are thereby implicitly naturalized as the true representatives of the community's best interests. Second, the very framing of the debate as a "classic environmental battle" reduces the local problem to a global abstraction that is irresolvable. No option for effective action exists. Someone is going to be hurt. Progress has its cost, and someone must pay. As the headline suggests, pollution is the "price" and most citizens are willing to pay that price for "prosperity." The article begins this way:

> The first thing you see of this small but busy little burg is the smoke. Viewed from faraway Interstate 24, it rises in bursts of gray at the horizon, sending mixed signals from deep within the western Kentucky lowlands.
>
> One message is that Calvert City (pop. 3,004) is hard at work. Eight sprawling factories and 2,300 employees are churning out ingredients for plastics, preservatives and other products. As in the Pittsburgh of previous decades, the smoke here means a robust economy and high-paying jobs.
>
> But for some area residents, it is a reminder of other things, too. A three-square-mile piece of Calvert City emits more toxic air pollution than the entire state of Colorado, accommodates millions of pounds of dangerous chemicals and lies above at least 11 hazardous waste sites.
>
> The divergent views are the touchstones of a classic environmental battle unfolding in this community. . . . (Charlier A5)

Beginning with the memorable image of the smoke and the varying interpretations that attach to it as either a sign of hard work or a sign of danger, this passage establishes a structure that will be repeated throughout the article. An interchange, a debate in the third person, is set up between the developmentalist perspective and the environmentalist perspective. The reporter works hard to present the case of each side. On the developmentalist side, there is this: "The total payroll [of the local manufacturers] is $100 million, and the manufacturing workers, earning an average of $620 a week, are among the best-paid in western Kentucky, according to 1985 figures" (Charlier A5). We are also told, however, that "the economic success has come at some ecological cost. According to a 1984 study, the industrial complex accounts for a staggering seventy-six percent or more of all the hazardous waste generated in the TVA region—although most of it is dilute and relatively innocuous wastewater treated on-site. Groundwater beneath four companies has been contaminated" (A5). In this passage, there is some evidence of either an unintentional effort to undermine the arguments or simple confusion. The qualifying clause about the innocuousness of the wastewater in the second sentence sits uncomfortably with the claim in the last sentence about groundwater contamination. Most likely, this non sequitur results from confusion, since the article continues with an impressively full account of the ecological problems: "Six of the plants released a total of 19.5 million pounds of toxic waste in 1987. . . . The amount of airborne toxic waste—10.5 million pounds—exceeds the totals for 11 individual states. Those air pollutants include 1.29 million pounds of 11 compounds EPA knows or believes to cause cancer in humans. . . . The plants also release 28 tons of carbon tetrachloride, which besides being a suspected carcinogen destroys protective ozone in the upper atmosphere" (A5).

If any perspective is favored by the length of evidence presented, it is the environmentalist side. This advantage may be traced, however, to the ostensible motive of the story—a recent EPA investigation into the ecological consequences of the industrial emissions in the area. The figures and technical data, as in any government study, are plentifully there for the taking. Moreover, the job and salary figures have such an impact that any elaboration of the developmentalist perspective is unnecessary. Though the meaning of the statistics on the millions of pounds of toxins released in the air will be lost on most readers, the average weekly salary figure, an ample $620, will speak most eloquently.

To detract further from the meaningfulness of the statistics on emissions, the reporter wraps the figures in controversy. Do the emissions cause cancer or not? "Brandishing maps with dots showing the homes of cancer victims," the environmentalists say yes. "However, health data compiled by local, state and federal officials has failed to identify any unusual cancer threat associated with the emissions." In fact, "when the figures are adjusted for the age spread, the county's cancer death rate actually is below that for Kentucky and the rest of the nation." The state epidemiologist says "it's unlikely—but not impossible—that the emissions are affecting residents' health." Predictably—and from the environmentalist perspective, devastatingly—the state scientist says, "we don't have all the information we need" (Charlier A5).

Hidden Agendas, Masked Conflicts

A yet more subtle rhetoric covertly operates in the portrayal of the environmentalist resistance. Consider this passage from the *Commercial-Appeal* article:

> Brandishing maps with dots showing the homes of cancer victims, citizens' groups have called for the shutdown of a commercial waste incinerator in the town and other sweeping measures.
>
> "We got a terrible thing here—it's almost criminal," says B. N. Dossett, a 75-year-old state retiree living in Calvert City. "The town was developed for the benefit of the few and no regard at all for the health of the many."
>
> Civic leaders, employees and some townspeople have responded by rallying to the support of the plants, arguing that the environmentalists' charges are unfounded.
>
> "They (the companies) are good corporate citizens for this region, and they always have been," says Carol Rogers, public relations and economic development director for the eight-county Purchase Area Development District. (Charlier A5)

One side is portrayed as "environmentalists" and "citizens' groups" (a phrase resonating lightly with hints about minority politics and special interests). The other side is said to be populated by "civic leaders," "employees," and "townspeople," the economically productive members of the community. To reinforce the minority status of the environmentalists, their ages are given every time they are quoted, or they are reported to be retired and are thereby associated with the "large elderly population" that is drawn to the area by "the nearby lakes and low cost of living."

Beneath the environmentalist/developmentalist conflict, then, lies the struggle between the working community and the residents that have flocked to "one of the nation's prime retirement locations." The ages of the developmentalists are never given, even though some of them, we can infer, are quite old: " 'We moved here because of them (the plants), and you can look at us. We look pretty healthy,' said Dortha Mathis, whose husband retired after 36 years at the plant and whose sons work there now." An insidious contrast thus emerges between the productive citizens of the region who came to work and the retired folks who came to enjoy the lake and complain about the air quality. We are in the presence of a media stereotype that has haunted environmentalists—the image of "little old ladies in tennis shoes," the "blue-haired brigade" with nothing better to do than to protest against the conditions that result from honest work by the productive and dominant forces of the middle-aged middle class.

This theme predominates right to the conclusion of the story, which at first seems to be an odd anticlimax to a highly dramatic report. The final word on the situation is given to an elderly man whose comments are set in opposition to Dortha Mathis' favorable impression of the industries:

> But Ed Dismore, 83, who lives next to the railroad tracks leading to the industries isn't so sure. "There's rank stuff that goes by here every day. . . . One of them suckers bust and it's goodbye, Irene," he said.
>
> Dismore, who spent much of his life in Missouri and other parts of Kentucky, believes Calvert City and other parts of the United States have become plagued by toxic air pollution. He is fatalistic about the issue, however.
>
> "What's there for me to worry about?" he says. "I'm too old." (Charlier A5)

Though seemingly inappropriate, the anticlimax is in fact right on the mark. The old man's fatalism is shared by the article itself. The conflicts that are reported are irresolvable. The struggle between environmentalist and developmentalist is itself naturalized, made universal by its parallel relation to the ancient conflict between old age and middle age. We pay the price of youth and middle age in our last years. And just as certainly, the benefits of industrial growth will exact their dues as industries mature in a community. No change is in sight. No change is even thinkable.

This same story has been told over and over again in various settings around the United States. We have heard about the conflict of labor and environmental protection in the lumber industry in California and Oregon, the mining industry in New Mexico and Utah, the defense industry in Nevada and Texas, the paper mills in the Carolinas. Insofar as the papers and newsweeklies can be counted on to favor short-term economic arguments—middle-aged, middle-class industrialism—they can be said to do the ideological work of the developmentalist perspective. Action groups, special interest organizations of outsiders like the Sierra Club and Earth First!, must take the role of the adversary, which has historically been the position assigned to the environmentalist. Like senior citizens and other minority action groups, such cadres of organized resistance enter the debate from the margins of society or are marginalized by the opposition.

Wilderness Protection and the Image of the Outsider

By their very name, environmentalists are identified as *outsiders* through their association with the environment, the external, the out-of-doors, that which surrounds, the margins of civilization, and, above all, *the wilderness*, to whose protection the first environmentalist groups, the Sierra Club and the Wilderness Society, were devoted. As the cultural anthropologist Linda Graber has suggested, "the wilderness ethic," the motivating ideology of environmentalist interest groups, "is primarily a philosophy based on the avoidance of man-caused enviornmental changes, so the application of wilderness criteria to questions of environmental quality in settled landscapes tends to force purists into a negative and reactive political stance" (114).

In *Wilderness and the American Mind*, the intellectual historian Roderick Nash traces the prehistoric, psychological roots of the wilderness concept to the emergence of agriculture as a social pattern in human life. *Wilderness* means land that is not cultivated or otherwise under civilized human control. It is the province of wild beasts and savage native tribes (Nash xiii). Along the same lines, René Dubos notes in *Wooing of the Earth* that "the word wilderness occurs approximately three hundred times in the Bible, and all its meanings are derogatory" (10); in traditional Western thought, wilderness is the enemy of civilization: "Humankind has struggled against environments to which it could not readily adapt" (11). Dubos further suggests that the "nature religion" associated with environmentalism draws converts mainly from among urbanized intellectuals, long uprooted from their agricultural heritage (13). Graber thus characterizes the wilderness as the "Wholly Other" to which modern mystics have turned for spiritual refreshment and re-creation: "The ancient Eden represented pre-industrial man's need for a vision of ease and rustic comfort to sustain him through a life of backbreaking work. Affluent Americans have achieved physical comfort in their daily lives, so perhaps they are redefining the Edenic landscape to include challenge as well as purity" (Graber 2, 15). Cultures that sustain a memory of their hunter-gatherer origins have no word for wilderness. As the Sioux chief Standing Bear wrote, "Only to the white man was nature a wilderness." For nomadic peoples like the Sioux, "it made no sense to distinguish wilderness from civilization or wild animals from tame ones": "Civilization created wilderness" (Nash xiii–xiv).

As early as the 1850s, transcendentalists like Henry David Thoreau, poets like William Cullen Bryant, and naturalists like George Perkins Marsh called for the preservation of wild places against the intrusion of civilization. Their arguments were commonly based on the aesthetics of the sublime and other neo-romantic concepts. In his Harvard commencement address, for example, Thoreau claimed, "This curious world we inhabit is more wonderful than it is convenient; more beautiful than it is useful; it is more to be admired and enjoyed than used" (qtd. in Brooks xiii). Early advocates of wilderness preservation were not averse, however, to developing utilitarian arguments designed to win over their more practically minded readers. Marsh was particularly keen on arguing that the continued existence of wilderness ensured the integrity of the American soil and water reserves. He showed how the clear-cutting of forests on the watersheds of rivers, for example, could lead to drought, flood, and erosion. "Because it made protecting the wilderness compatible with progress and economic welfare," writes Nash, "Marsh's argument became a staple for preservationists" (105). Yet another staple, which has remained with the environmentalist cause to this day, was the claim that governmental regulation was needed to ensure the preservation of wilderness (Nash 105).

The Emergence of Conservation and Reform Environmentalism

The moving spirit of the preservationists was John Muir, the essayist-geographer-geologist-mountaineer who was instrumental in the preservation of the Yosemite and who founded the Sierra Club in 1892, when the age of reform was at its height. In his early political battles, Muir called himself a "conservationist" and did not hesitate to use utilitarian arguments. Anxiety over the realization that the abundant natural resources of North America could actually be depleted drove a number of practical men and women into the same camp with transcendental nature mystics like Muir himself. This coalition of conservationists found a common enemy in those who unwisely used unprotected resources.

Eventually, though, it became clear that a "schism ran between those who defined conservation as the wise use or planned development of resources and those who have been termed preservationists, with their rejection of utilitarianism and advocacy of nature unaltered by man" (Nash 129). Muir's adversary in the conflict, Gifford Pinchot, eventually "succeeded in appropriating the term 'conservation' for the wise-use viewpoint," while Muir and his frustrated followers, now denied their name of choice, lamely denounced Pinchot's camp as "de-conservationists" (Nash 139). Ultimately winning the unflagging support, indeed the total devotion, of President Theodore Roosevelt, Pinchot cultivated in his newly developed Department of Forestry the "gospel of efficiency," applying the concepts of scientific management in a program of wise use of resources. In this approach, Pinchot established the government position that endures to this day, mediating between unbridled capitalist development and uncompromising demands for wilderness preservation.[1]

Muir's political descendants are the environmentalist interest groups. In opposition to both the government-sponsored wise-use version of conservationism and the capitalist "exploiters" of natural resources, wilderness activists have played the role of the purist, not averse to modeling right action and wise use, but reaching beyond anthropocentrism to the position of the nature mystic, who loses self in the contemplation of a higher reality—the Wholly, and Holy, Other. Muir himself was known for his rapturous nature worship as well as his fights for governmental protection. Growing up in a strict Scottish household ruled by a domineering and puritanical father, Muir learned little about natural religion, but he must have developed there much of his rhetorical flair. In his later years, he was not above associating his work with the name of God. Thus Thoreau's famous sentence, "In wildness is the preservation of the world," Muir rendered as "In God's wildness lies the hope of the world" (Brooks 19).

Nevertheless, when it came to the preservation of "the remnant of our forests," Muir was a realist who believed that the federal government might do more than God: "Through all the wonderful, eventful centuries since Christ's time—and long before that—God cared for these trees, saved them from drought, disease, avalanches, and a thousand straining, leveling tempests and floods; but he cannot save them from fools,—only Uncle Sam can do that" (Muir 364–65).

Since World War II, Muir's descendants in the Sierra Club have opted to follow the track of political realism, leaving the transcendentalist strain of Muir's work to the deep ecologists and the new agers, though reform environmentalists will still occasionally add a high tone to their rhetoric in the tradition established by Muir. According to Graber's analysis, they mainly use religion "as a metaphor rather than as an explanation," giving their polemics titles like "Kick the Exploiters Out of the Wilderness Temple" (8). On this basis, David Brower, the former executive director of the Sierra Club, is compared to the evangelist Billy Graham in John McPhee's *Encounters with the Archdruid* (83). On the whole, however, one can hardly imagine a more secular organization than the Sierra Club. It has preserved Muir's memory mainly in its dual emphasis on wilderness aesthetics and political action. Its book catalogs may feature on facing pages collections of Ansel Adams' sublime photography of the Yosemite—the effect of which on the development of environmental consciousness should not be underrated, as Samuel Hays has noted (*Beauty*

37)—and Wendell Berry's cranky tract on the corruption of American agriculture, *The Unsettling of America*. The activities of the club's local meetings likewise range from members' slide shows of their recent canoe trips to planning sessions for political resistance against the pressures of developmentalism. The Sierra Club *Bulletin* features both the journals of mountain climbers and the highly charged editorial outbursts of environmentalist militancy.

The fervor of the Sierra Club's militancy is directed against a well-defined image portraying a diffuse, even all-pervasive enemy, the "exploiters" in their many guises—government conservation experts, the Army Corps of Engineers, land-hungry real estate developers, oil and mineral companies. The standard tactic of Sierra Club rhetoric is to subvert or invert the developmentalist claim that growth, especially economic growth, is always good. Against this line of reasoning, it is asserted that some kinds of growth are bad and that furthermore, even when growth is good in the short run, it may yield evil in the long run. The growth mentality, which promises the good life to local inhabitants and in reality promotes the interests only of the powerful few, may prove as unnecessary for the general public as it is damaging to the land. In a 1971 contribution to the club's *Bulletin*, Edward Abbey develops a characteristic analogy between the ideology of growth and a cancerous disease: "Like the network of new highways proposed for the canyon country, these power plants are meant not for current needs but for 'anticipated' needs. 'Planning for growth,' it's called. The fact that planning for growth encourages growth, even forces growth, would not be seen as a serious objection by the majority of Utah-Arizona businessmen and government planners. It would merely excite them to greater enthusiasm. They believe in growth. Why? Ask any cancer cell why it believes in growth" ("Canyonlands and Compromises" 393).

In a 1947 piece in the *Bulletin*, Bernard DeVoto draws on the century-old conservationist argument and adds an apocalyptic touch to his characterization of the foe: "If the watersheds go, and they will go if cattlemen and sheepmen are allowed to get rid of government regulation of grazing, the West will go too—farms, ranches, towns, cities, irrigating systems, power plants, business in general. Much of the interior West will become uninhabitable, far more will be permanently depressed" (391). And Garrett Hardin, in a *Bulletin* editorial reprinted from *Science*, readily accepts the pejorative labels offered by the critics of the Sierra Club in a scathing dismissal of industrialist growth in the face of overpopulation:

> Those of us who are deeply concerned about population and the environment—"econuts," we're called—are accused of seeing herbicides in trees, pollution in running brooks, radiation in rocks, and overpopulation everywhere. There is merit in the accusation. . . .
>
> People are dying now of respiratory diseases in Tokyo, Birmingham, and Gary, because of the "need" for more industry. The "need" for more food justifies overfertilization of the land, leading to eutrophication of the waters, and lessened fish production—which leads to more "need" for food. ("Nobody Ever Dies of Overpopulation" 486–87)

Members of the Sierra Club were among the first citizens to accept the name "environmentalist" as their own, to present themselves publicly as environmentalists (see Ela's 1972 article, for example). As Nash notes, as late as 1960 the Sierra Club was "a cozy, largely regional organization of 15,000 mountaineers," but by 1967 "it was on its way toward 60,000 members and recognition as a major national force in the newly launched 'environmental' movement" (ix).

Despite its growth in membership, the Club retained, in the late 1960s and early 1970s, its mountaineer aesthetics and politics, due to which it remained a limited countercultural force; still it represented the outsider and was thus all too vulnerable to the attacks of developmentalists, especially those who claimed that the environmentalists were elitists more concerned with their own petty pleasures over a beautiful vista—a "scenic climax," as David Brower liked to say—than with the survival of the great masses of American society.

Environmentalism under Fire

When James Watt gained the public's attention, first in the late 1970s as the leader of the Mountain States Legal Foundation and then in the early 1980s as President Reagan's Secretary of the Interior, he directed his arguments precisely against the mountaineer wilderness ethic in an attempt to reinforce minority status, the otherness of "special interest groups" like the Sierra Club. He would claim that the self-interested environmentalists were really out to preserve hiking spots inaccessible to many Americans. In a 1981 interview, he asserted that "America's resources were put here for the enjoyment and use of people, now and in the future, and should not be denied to the people by elitist groups" (Marth 36). He specifically named the Sierra Club as one of these groups and ignoring the environmentalists' purer interests in wilderness for its own sake, attacked the wilderness ethic as motivated by the mountaineer's desire for remote vacation lands: "I believe that these [public] lands . . . must be made available to the handicapped, to the family, to the camper, to the older couple, to the young couple with not so much money. I do not think they ought to be set aside for just the few backpackers who are rugged enough and affluent enough and have enough time" (Marth 39).[2] Watt thus attempted to undermine the influence of environmentalism by creating a conception of the environmentalist character that would repel the ordinary citizen. Much as in the covert rhetoric of the newspaper report we analyzed earlier in this chapter, Watt's comments emphasize the gap between working people who just want a good job and environmentalist intellectuals who, with relatively secure employment for themselves in professional and managerial positions, are out to make it hard for other people to get work. In an early 1983 interview, the same tactic emerged with renewed force: "When we campaigned in a university town with high employment, we had ugly demonstrations. When we went into a community with high unemployment we had no ugly demonstrations. . . . When there are two lines and one is an ugly group of elitists demonstrating against America and another group trying to seek jobs, I'll go help the job searchers" ("Secretary Watt" 86). Significantly, Watt identifies himself with the nation as a whole in this passage, suggesting that the intellectuals were demonstrating not against his own policies but "against America." Not satisfied with characterizing environmentalists as elitists, Watt would portray them as "anti-American" and would go so far as to stigmatize them as communists and enemies of democracy: "Their real thrust is not clean air, or clean water, or parks, or wildlife but the form of government under which America will live. The environment is a good vehicle to achieve their objectives. That is why you see the hard-line left drifting toward that interest" ("Secretary Watt" 86).

For all his opposition to the Sierra Club, Watt used many of the same rhetorical techniques. Some of his supporters in the oil industry told us that he consciously determined to fight fire with fire in the late 1970s when the environmentalist lobby had reached its first peak of success (Killingsworth, "Can an English Teacher"). It was at this time that he developed a series of apocalyptic tales of the future to counter the tales told by the environmentalists. Here is one sample: "We do not have an energy crisis; we have a crisis in government management. If we do not allow the private marketplace to go in and develop these energy sources in a systematic, methodical, and environmentally sensitive way, we will create such a political and economic crisis that Washington will nationalize the industries and attack our energy-rich West in such a manner as to destroy the ecology, primarily because it must get to that energy to heat the homes of the Northeast and keep the wheels of industry going in the Midwest" (qtd. in Marth 40). Once again we see the rhetorical strategy of overlaying on the developmentalist/environmentalist conflict another more troubling and deeply rooted conflict—the struggle of capitalistic democracy to maintain its goals against socialistic intrusions. If Watt's critique of the "elitist" wilderness ethic, his championing of open access to public lands, and his recognition of the growing interest in environmental amenities led to his ascension to power in the popular Reagan administration, his ideological intensity may have proved a bit too much for the general public, who eventually came to view him as yet another marginal character, just as much out of the mainstream as his opponents in the Sierra Club.

The Extension of the Environmentalist Ethos

The fears of Watt and his conservative colleagues must have abated somewhat with the passing of the energy crisis and with the setback the environmental movement suffered when the Reagan administration came to power. But Watt's failure to build a political coalition to support his developmentalist program reflected, as Hays suggests, "a lack of public support for the administration's policies" (*Beauty* 514). The Reagan presidency, which "set out to undo the environmental work of the preceding two decades of Republican and Democratic leadership," finally was "forced to recognize that environmental affairs were not momentary, limited, and superficial"; ironically, the actions of this administration ultimately "strengthened public support for environmental organizations" (*Beauty*, 491–92, 505).

The general support of environmental values was consequently reflected strongly in the presidential campaign in 1988. Republican George Bush, an avid outdoorsman who had been Reagan's vice president and a Texas politician with long-standing oil company connections, proclaimed, "I am an environmentalist" and attacked the environmental record of his Democratic opponent. It would have been more in character if Bush had said "conservationist," an advocate of government control of natural resources, but his acceptance of the environmentalist label has a potentially deep significance—even if, as skeptical observers suggest, the rhetoric is shrewder than it is substantial. For even if Bush's actions don't fulfill his promise to make his administration "the environmental presidency," he has created a noteworthy moment in rhetorical history by certifying environmentalist values as a valid component of the presidential ethos as it reflects the American character in general. Indeed, as the presidential race heated up during the unusually warm 1988 summer, the New York *Times* reported a "surge of support for the environmental movement," prompted by concern over the degradation of shorelines and the fear that "pollution may be altering the climate"; this fear prompted a significant increase in monetary contributions to environmental study and action groups (May 20Y).

As in the time of the energy crisis, historical, external events may be said to have turned the public's head more than the rhetoric of the environmentalist polemics that James Watt geared up to fight. Polemics reflect history as much as they shape it. It is no accident that the Sierra Club was founded in the 1890s; when city life was becoming increasingly intolerable because of the "environmental costs" of the industrial revolution, people sought solace in the wilderness and determined to preserve it against further encroachments. The development of wise-use policies and conservation science may have delayed the burgeoning of reform environmentalism for nearly a century. Along with the shifts of support for or against environmental protection in the late 1980s—shifts reflecting an acute perception of the new levels of environmental degradation—we are now witnessing some major new directions in the rhetoric of environmental politics. These alterations in discourse reflect the development of new perspectives on the problem of life in an ever more crowded, affluent, and polluted world.

Shaping and Controlling the Discourse of Environmentalism

If everyone from a Sierra Club activist to President Bush has some claim to the title *environmentalist*, we may well wonder whether the word has been extended to the point of meaninglessness. As Neil Evernden suggests, "The term 'environmentalist' was not [originally] chosen by the individuals so described. It was seized upon by members of the popular press as a means of labelling a newly prominent segment of society." Nevertheless, Evernden continues, "any term selected by the news media is likely to be drawn from common usage and to reflect common assumptions" (125). Indeed, the sharing of the term may finally bring to our attention what has been hard for many social observers to grasp: Environmentalists hold in common with their ostensible opponents a number of values, practices, and elements of discourse. They tend, for example, to express their appreciation for the land in terms of possession, consumption, or personal experience. The land is a treasure, a source of aesthetic delight with mystical properties, a "scenic orgasm." The land may be used as a sink for industrial waste, or it may be used for resort retirement property, prime backpacking country, a preserve for endangered species, but in any event it is *used*. In light of the shared belief in use-value, the distinction between environmentalism and developmentalism may come to seem trivial, and ecospeak a mere diversion, the stuff of dramatic newscasts.

Environmentalism is an outgrowth of the general liberal temperament and ideology. Only recently—and with precious little success—have philosophers, green politicians, deep ecologists, and eco-anarchists sought to reach beyond this political framework to a new discourse and new forms of action.

The Word

Despite its present status as a household word and its currency in the mass media, *environmentalist* is not an old word nor does it have a distinguished history. As a designation of "one who is concerned with the preservation of the environment (from pollution, etc.)," it is barely two decades old. The first instances of this usage cited in the *Oxford English Dictionary* (1972 Supplement) are from 1970: "Dr. Robert N. Rickles, a thirty-four-year-old chemist and an *environmentalist*, . . . took the helm of the city's Department of Air Resources last month" (*New Yorker* 9 May 1970; italics added); and "the project to build a supersonic transport has run into renewed complaints from the *environmentalists*" (*Nature* 15 Aug. 1970; italics added). The definition of the term implied in these samples (one in an American magazine, the other in a British journal) was included in *Webster's Third International Dictionary* only as recently as the 1986 edition, where it appears in the addenda.

The definition is matched in the *OED* citation with another usage: "One who believes in or promotes the principles of environmentalism," the "theory of the primary influence of environment on the development of a person or group." This meaning of the term in natural science applies to evolutionary theorists in the Lamarckian tradition, who have asserted, against the more purely Darwinian geneticists, that factors external to an organism may leave their traces and influences in the genetic history of the organism's descendants. By the mid-twentieth century, this view had fallen into disrepute, and *environmentalism* had become a term of derision attached to those outside the mainstream of scientific research.

Both the popular and the scientific meanings of *environmentalist* were originally used not as self-designations by proponents of environmental protection or evolutionary environmentalism but were applied by opponents or observers of these proponents. The term began in both instances as a designation for outsiders. Whereas in the scientific usage *environmentalism* denotes an opposition to the accepted dogma of geneticism, however, in the popular usage the term exists without a clear opposite, suggesting an ideological content, a distinction between the environmentalists and the rest of us. What would the opposite be? "Capitalist" is a likely candidate, for capitalism is based upon the pursuit of profits in an endless upward spiral regardless of the constraints posed by the environment or by labor. But not everyone who opposes environmental protection is a capitalist. The commitment of organized labor and community developers to providing the maximum number of jobs for workers has set working people against environmentalists in many regions. Thus in our previous analysis, we have had to coin the makeshift term *developmentalism* to describe the opposition to environmentalism. We are fully aware, though, that this practice is misleading, for it suggests that the meaning of *environmentalist* is more settled than it really is. In truth, the term has not reached the place in its history where it can be said to have a clear opposite.

Though George Bush and a growing number of Americans have recently laid claim to the word, a number of writers that most readers would place among the leaders of environmentalist thought, research, and activism, would totally reject the label. Indeed for them, the very word *environment* is questionable. The influential poet and cultural critic Wendell Berry, for example, has written, "The concept of country, homeland, dwelling place becomes simplified as 'the environment'—that is, what surrounds us. Once we see our place, our part of the world, as surrounding us, we have already made a profound distinction between it and ourselves" (22). According to Berry, a particular perspective—the perspective of a humankind alienated from the land—is implied in the very term that predominates in the discussion of present-day humanity's relationship to nature. Implicit in Berry's preference for "country" or "homeland" and his criticism of the alienated view of the "environment" is a remedial view, the perspective of a people at home in the natural region on which their sense of identity depends.

From the perspective against which Berry sets himself, however, this longing after homeland may appear nostalgic, even reactionary. The comfortably alienated technologist—or George Bush himself, who left his New England home to make his fortune and begin his political career in Texas—might claim that the concept of homeland as a source of identity is outmoded in our mobile and cosmopolitan society, that we have been freed from dependence on the land for a sense of identity. The environment is merely what surrounds us from day to day, not some quasi-magical entity on which our very selfhood depends. Moreover, the argument might run, the word *homeland* has unfavorable associations with fascist nationalism.[3]

Ironically, there remains a curious dependence of each side upon the other and an interesting likeness in the rhetorics. Though Berry may identify himself with the land and thus may issue a call for unity with nature, he can at best hope for *unification* as a goal for the audience he addresses. These readers, he feels, are in need of his perspective to remediate their own disunity; they in turn may respond positively to his discourse not out of genuine identification with the earth, lacking as they do a history of close associations with the land, but rather out of an aesthetic impulse to preserve the pastoral landscape and way of life in the face of vaguely perceived forces that threaten their own existence as surely as they threaten the family farm. Likewise, George Bush cannot hope to feel automatically accepted among his environmentalist opponents merely by saying, "I am like you"; more likely his self-designation as an environmentalist represents an effort to appeal to a third party, the proverbial average voter, who does not identify strongly with the environmentalist cause as such but feels a definite longing for better surroundings. As Kenneth Burke notes, " 'unification' is not unity, but a compensation for disunity" (*Grammar of Motives* 173).

Thus, although the very use of the term *environment* implies an acceptance of human alienation from nature, a critique of usage on these grounds implies a similar acceptance of the alienated condition.[4] Moreover, the striving for unity with the environment often leads to disunity with one's fellow citizens, who are, though alienated from the land, necessarily part of one's "dwelling," the "social environment" in which one lives: Wendell Berry lives on a farm in Kentucky surrounded by an American public whose contrasting ways of life contribute to his own self-definition as a radical ecologist, poet, and farmer. The irony in this position is not lost on opponents of environmental protection who are quick to claim that environmentalists care more for trees than for their fellow human beings. The real estate investor and developer Charles Fraser—one of the antagonists in *Encounters with the Archdruid*, John McPhee's book about the Sierra Club activist David Brower—takes the old line of reasoning in calling himself a conservationist who stops short of the preservationist efforts of most environmentalists. Preservationists, he says, neglect human needs. McPhee quotes him as saying, "Ancient Druids used to sacrifice human beings under oak trees. Modern druids worship trees and sacrifice human beings to those trees" (95).

Very likely, the common division of the "social environment" from the "natural environment" in environmental impact assessment is likewise based on this deeply felt perception of human needs in conflict with the needs of nonhuman nature. Of course, the "druids" themselves claim to have humanist ideals; in their offering to nature, they hope to prepare a fit place for the reproduction of human life. Questions of the "short run" versus the "long run" and of whose interests are most clearly directed to the happiness and preservation of human beings—the short-term focus of developmentalism with its offer of better jobs and better housing or the long-term concerns of environmentalism with its questions about the future we are making for our children and their children—thus founder on a division of perspectives that seems irreconcilable.

How to refer to one's own position in the environmental debate is therefore among the leading semantic and pragmatic problems writers face. To "refer to" is in fact a large part of how to *construct* a position on environmental issues.

Ecology

Like *environment*, *ecology* is a term that frequently rises to a position of privilege in the language of the avant-garde in environmental politics. Possessed of a homely etymology—derived from the Greek word for *household* and coined in 1870 by the German biologist Ernest Haeckel—the word was gradually adopted by biologists as the term for the study of "communities," organisms in their relation to each other and to their surroundings. It was taken up one hundred years after its invention as a slogan by political environmentalists who sought to limit technological development and to reconstruct the human relationship with other members of the biotic community. In the early 1970s, the obscure phrase "Ecology Now!" was printed on bumper stickers along with a green theta, the Greek letter that originally symbolized life. (Recently the theta has been revived on a bumper sticker that no longer uses the word "ecology," but offers the message, "BACK BY POPULAR DEMAND.")

In a discussion of the painful awakening from the "enchantment with technology," the philosopher and cultural historian Thomas Berry adopts the term *ecology* as a designation of the new age of enlightenment that he foresees:

> Presently we are entering another historical period, one that might be designated as the ecological age. I use the term *ecological* in its primary meaning as the relation of an organism to its environment, but also as an indication of the interdependence of all the living and nonliving systems of the earth. This vision of a planet integral with itself throughout its spatial extent and its evolutionary sequence is of primary importance if we are to have . . . transformations [that] require the assistance of the entire planet, not merely the forces available to the human. Otherwise we mistake the order of magnitude of the challenge. It is not simply adaptation to a reduced supply of fuels or to some modification in our system of social or economic controls. Nor is it some slight change in our educational system. . . . It is a radical change in our mode of consciousness. Our challenge is to create a new language, even a new sense of what it is to be human. It is to transcend not only national limitations, but even our species isolation, to enter into a larger community of living species. This brings about a completely new sense of reality and of value. (41–42)

To bring about this new transcendent language and spirit, Berry extends the "primary meaning" of *ecology* well beyond the boundaries of scientific usage but without rejecting the meaning of the term as it is used in science, a meaning that roughly corresponds to "the relation of an organism to its environment." In including "the interdependence of all the living and nonliving systems of the earth," moreover, Berry incorporates not only the general topic of scientific ecology but also one of its chief findings. This inclusion of science is crucial to his vision, which is reminiscent of the outlook of Emerson, Whitman, and Thoreau, the nineteenth-century poet-prophets who engulfed science in their transcendental quest for an image of completeness and unity. Berry claims, however, that the ecological viewpoint of the present age transcends even the transcendentalists; the new perspective, he argues, "was unthinkable in ages gone by" (42). "The ecological age into which we are presently moving is an opposed, though complementary, age that succeeds the technological age. In a deeper sense this new age takes us back to certain basic aspects of the universe which were evident to the human mind from its earliest period, but which have been further refined, observed, and scientifically stated in more recent centuries" (44). Despite his respect for the identity with the earth fostered in the cultures of Native Americans and other non-Western groups, Berry asserts that "nowhere was the full genetic relatedness of the universe presented with such clarity as by the scientists of the twentieth century" (46).

Not all activists are as sanguine about the accomplishments of modern science and are therefore slightly uncomfortable in associating their own perspectives with those of academic ecology. Jonathan Porritt, the leader of Britain's Ecology Party in the early 1980s, admits that " 'ecology' is still a rather daunting word, perhaps too scientific, too specialized to convey the full scope of the green perspective" (3). Porritt and other European Greens have come to believe that to align themselves with the science of ecology may represent a violation of their values, since science tends to rely on the same post-Baconian techniques of manipulation and prediction in its dealings with nature that characterize the actions of techno-capitalism. Thus science as currently practiced can only be opposed to green politics and can only, in the long run, depend upon the monetary and ideological support of capitalists and government conservationists. As Evernden points out, many scientific ecologists are equally uncomfortable with having been "linked in the public mind with a rag-tag collection of naturalists, poets, small-scale farmers, and birdwatchers" who have adopted the title of a scientific discipline as the name on their political banner (5).

Ignoring the semantic difficulties of sharing their central term with traditional scientists, some philosophers and social theorists have followed Arne Naess in adopting the term "deep ecology" primarily to distinguish themselves from reform environmentalism and to hint at the need, which they recognize in common with Thomas Berry, to restructure the very foundations of human culture in order to restore ecological balance. In *Deep Ecology: Living as if Nature Mattered*, Bill Devall and George Sessions write, "Reformist activists often feel trapped in the very political system they criticize. If they don't use the language of resource economists—language which converts ecology into 'input-output models,' forests into 'commodity production systems,' and which uses the metaphor of human economy in referring to Nature—then they are labeled as sentimental, irrational, or unrealistic" (3). Hence a language that constructs a new, deeper consciousness, a new outlook on the world, indeed a new world, is needed.

In light of their call for this bright new vision, however, it is ironic that the deep ecologists adopt a term from science to describe their movement, a term older than environmentalism itself. We may wonder, if reform environmentalism is "shallow," and if deep ecology is deep, what exactly is scientific ecology? In their semantic acrobatics, the deep ecologists have betrayed serious signs of what the literary critic Harold Bloom has called "the anxiety of influence." Despite their objections, they stand in a direct line of descent from both reform environmentalism, which they claim to have replaced, and the holistic branch of scientific ecology, which they either ignore or appropriate in ways that most scientists would find objectionable.

The rhetorical program of the deep ecologists is an important one, however. Their denial of their lineage is a rhetorical form of self-assertion, an effort to break the hold of ecospeak and to create a fresh discourse based on new alliances. To say, "We are not environmentalists!" is to reject many current forms of discussion. Deep ecology is thus the historical vanguard in the extension of environmentalism, an extension that corresponds to the growth of the relatively affluent urban classes and the spread of environmental pollution to ever wider regions of the earth, including even the soils, the air, and the oceans. More a world view than a program for action, the movement represents one version of a global environmental consciousness spawned by the startling findings of scientific research in the last half of this century and by the potentially radical premise of all versions of ecology that everything in nature is interrelated.

To gain a full appreciation of this step from local to global discourse, we must examine in greater detail the attempt to transform scientific discourse into public rhetoric. The connection between science and the environmental reform movements—a match directly encouraged by authors like Thomas Berry and implied in the perspective of deep ecology—has become the most problematical and the most important link in the evolution of environmental politics in America. The many uses of scientifically generated information form the topics of our next four chapters. Moreover, the place of science in environmental activism turns out to be, in our estimation, the crucial issue in the development of a renewed outlook.

Wisdom demands a new orientation of science and technology towards the organic, the gentle, the non-violent, the elegant and beautiful.

—E. F. Schumacher,
Small Is Beautiful (29)

Human nature exists and operates in an environment, and it is not "in" that environment as coins are in a box, but as a plant is in the sunlight and soil. It is of them, continuous with their energies, dependent upon their support, capable of increase only as it utilizes them, and as it gradually rebuilds from their crude indifference an environment genially civilized. Hence physics, chemistry, history, statistics, engineering science, are a part of disciplined moral knowledge so far as they enable us to understand the conditions and agencies through which man lives, and on account of which he executes his plans. Moral science is not something with a separate province. It is physical, biological and historic knowledge placed in a human context where it will illuminate and guide the actions of men.

—John Dewey,
Human Nature and Conduct (296)

The seeker after truth builds his hut close to the towering edifice of science in order to collaborate with it and to find protection. And he needs protection. For there are awful powers which continually press upon him, and which hold out against the "truth" of science "truths" fashioned in quite another way, bearing devices of the most heterogeneous character.

—Friedrich Nietzsche,
"On Truth and Falsity" (180)

2

The Rhetoric of Scientific Activism

Science and Ecological Ethics

To enhance the attractiveness of the environmentalist ethos and the wilderness ethic, the Sierra Club and other political action groups have always sought to demonstrate a scientific basis for their perspective. Writers in the Sierra Club *Bulletin* frequently appeal to science. "Management of our land must be based on biological principles," a typical article claims, "for ultimately these are the principles that determine our fate" (Hood and Morgan 405). The environmentalists have not had to look far for material to bolster their arguments. Recent findings on biodiversity and chemical cycles in the atmosphere, for example, have stimulated a strong revival of the old utilitarian argument about the usefulness of wilderness. As René Dubos, the champion of ecological holism, suggests, "Increasingly during recent years, interest in the wilderness and the desire to preserve as much of it as possible have been generated by an understanding of its ecological importance. It has been shown, for example, that the wilderness accounts for some 90 percent of the energy trapped from the sun by photosynthesis and therefore plays a crucial role in the global energy system" (*Wooing of the Earth* 16).

Even more important than the effort of environmentalists to assimilate scientific findings, however, has been the politically motivated writing of a few well-respected and talented representatives of the scientific community itself. These writers have brought the message of ecological holism to the public and have asserted the radical connection between science and social history in the face of strong resistance from the scientific community itself. The historical forces that shape discourse-practice in mainstream science present an obstacle to those who would assert scientific authority in matters of public policy; objectivism has created a politics out of the scientific majority's apolitical stance. According to the biology professor and environmentalist Barry Commoner, scientists have for the most part been "guided by the idea that it is the scientist's duty to pursue knowledge of nature for its own sake without regard to social consequences" (*Science and Survival* 104). The ultimate use of the research results is simply not an issue in this way of thinking. As we shall show in chapter 3, the scientist's insulation is built into the very structure of discourse in mainstream applied research, which though framed by human-oriented problems (such as pesticide pollution or wildlife habitat destruction), tends to concentrate on reporting data, offering conclusions that are shaped by scientific rather than public interests and risking few, if any, direct recommendations for public action. With this system in place, the scientist's desire for independence can be accommodated within the technological division of labor in such a way that both scientists and those who use scientific information for technological purposes can be satisfied. Many of the countercultural critics of modern technology have, on this basis, associated science with the developmentalist perspective and have suggested that the scientific ethic encourages acquiescence. However, as Commoner notes in his *Science and Survival* (1967), a "relatively new" ethical position has developed among scientists themselves, which asserts "a particular moral responsibility to counter the evil consequences" of scientific research, a responsibility deepened because scientists are "in possession of the relevant technical facts essential to an understanding of the major public issues which trouble the world" and are "trained to analyze the complex forces at work in such issues" (104).

In the writings of three environmental activists, all of them trained scientists, the alternative ethical voice has become increasingly strong over the last three decades. With each of these figures—Aldo Leopold in the 1950s (when *A Sand County Almanac*, published posthumously, had its first great impact), Rachel Carson in the 1960s, and Barry Commoner in the 1970s—the fears of the general public and the prophecies of the environmentalists have been substantiated and articulated with a newfound authority. In their work, a powerful irony engenders a strong tendency toward scientific self-critique. Commoner describes the goals of the scientific activist in this way: "By applying the traditional skepticism of science to science itself, we may find clues to the surprising failures of some of the new large-scale technological applications" (*Science and Survival* 48).

In Leopold's essays, one perspective slides ironically over another, giving rise to an originally scientific/objectivist recognition of the wilderness' otherness that ultimately takes on aspects of the old mysticism of Muir, moving in the direction of deep ecology and all but eliminating the distance between the two perspectives as we mapped them in our introduction. In Carson's *Silent Spring*, scientific facts and findings are presented to the public with a literary flare that dramatizes the appeal of ecological holism as never before. In Commoner's books, the scientific self-critique encourages a new cooperative venture between natural scientists and social scientists.

With varying degrees of trepidation, these writers cross over from the scientific *is* to the ethical *ought*. They all stand back from the localized research of the specialist, the typical endeavor of most working scientists, and take a global view of the ecological condition of the earth. From irony—a global questioning of policy in light of scientific findings—they move to a full acceptance of holistic environmentalism as their leading perspective on the state of the planet. In examining one at a time their contributions to the environmentalist canon, let us ask the question, What is gained and what is lost in this revised view?

Aldo Leopold and the Roots of Deep Ecology

The first important progenitor of the current ecological awareness movements was the naturalist and forester Aldo Leopold. Leopold's writing is a lesson in the rhetorical difficulties of putting forward a program for general ecological reform, which though grounded in his experience as a scientist and resource manager in the tradition of Pinchot, involved a transformation of scientific objectivism that would accommodate a personal inclination toward biocentric ethics and ultimately toward nature mysticism. His undeniable ability as a rhetorician could not altogether compensate for the semantic-political problems inherent in his critique of environmental policy in the period just after World War II. He nevertheless developed a rhetoric whose breadth of appeal is unsurpassed in the discourse of environmentalism.

"The Land Ethic"

In "The Land Ethic," Leopold's most famous essay, published in 1949 as a chapter in *A Sand County Almanac*, the scientist-manager shows a clear understanding of the semantic dimensions of the conservation problem (which for him included the concerns of *environmentalism*, two decades before there was even such a word). Acknowledging the need to rework crucial metaphors in order to achieve a new way of seeing our relation to nature, he writes, for example: "The image commonly employed in conservation education is 'the balance of nature.' For reasons too lengthy to detail here, this figure of speech fails to describe accurately what little we know about the land mechanism. A much truer image is the one employed in ecology: the biotic pyramid" (251).

When it comes to developing his central argument, Leopold again finds current language—and current ethics—inadequate. His thesis is developed in three characteristically terse but highly suggestive paragraphs:

> The first ethics dealt with the relation between individuals; the Mosaic Decalogue is an example. Later accretions dealt with the relation between the individual and society. The Golden Rule tries to integrate the individual to society; democracy to integrate social organization to the individual.
>
> There is as yet no ethic dealing with man's relationship to land and to the animals and plants which grow upon it. Land, like Odysseus's slave-girls [who were hanged because their master suspected that they had misbehaved during his absence], is still property. The land-relation is still strictly economic, entailing privileges but not obligations.
>
> The extension of ethics to this third element in human environment is, if I read the evidence correctly, an evolutionary possibility and an ecological necessity. It is the third step in a sequence. The first two have already been taken. Individual thinkers since the days of Ezekiel and Isaiah have asserted that the despoilation of the land

> is not only inexpedient but wrong. Society, however, has not yet affirmed their belief. I regard the present conservation movement as the embryo of such an affirmation. (238–39)

The rhetoric of this passage is subtle yet forceful, particularly remarkable for its appeal to the reader's democratic sense of progress. We have passed beyond the primitive ethics of ancient Greece, of the ancient Hebrews and early Christians, the argument runs; let us now go even farther and extend our forebears' democratic ethic to the land and to all of the earth's creatures. The conservation movement is figured as an "embryo," the development of the land ethic as an evolutionary process of natural growth, as if our only concern were to stand aside and let nature take its course. The implicit appeal to nature and to progress as a natural and inevitable process of growth is among the most effective devices of American middle-class rhetoric; Leopold knows his audience.[1]

Leopold's name for his new ethical perspective, the "land ethic," has not caught on, however—and for good reason. For he expects the little Anglo-Saxon noun *land* to carry too great of a semantic burden in an age that inscribes authority in impressive technocratic jargon. In contrast, and in a bold attempt to control the direction of environmentalist discourse, to turn it away from the province of the experts with their imposing technical language, Leopold makes a plea for an almost primitivistic simplicity: "The land ethic simply enlarges the boundaries of the community to include soils, waters, plants, and animals, or collectively: the land" (239). "The land is not merely soil" (255), he tells us, knowing full well that modern usage restricts the meaning of the term to just that. Though the word still designates the region on which a people depends for national, cultural, or racial identity, few people recall that denotation without an effort, though they may understand it in the context of folk songs ("This land is your land, / This land is my land. . . .") and readings in the King James Bible, to which Leopold alludes.

For most people now, *environment* has taken on the broad meaning Leopold hoped to reserve for *land*, perhaps because *environment* has a value-neutral tone more likely to appeal to a general audience that seeks legitimation in the discourses of the natural and human sciences.

In addition to its association with soil, *land* means *property*. Leopold feared this semantic development and the thinking that undergirds it, and in "The Land Ethic" he took measures to alter the trend. In all dealings with the land, there are two groups of people, he argued: "one group . . . regards the land as soil, and its function as commodity-production; another group . . . regards the land as a biota, and its function as something broader. How much broader is admittedly in a state of doubt and confusion" (258–59). These two groups create conflicting perspectives of the relationship between human beings and the natural environment: "land the slave and servant *versus* land the collective organism" (261). Leopold demurs when it comes to naming the two groups. But the group that regards the land as a slave is described in terms that clearly suggest the Marxian critique of capitalism as an outlook on the world that, through reification, levels all beings and conditions to a commodity value, puts a price on the land (and even on the people who labor on the land). Leopold avoids designations like "capitalist," though, presumably for rhetorical reasons, to avoid alienating the great majority of the potential constituency he addresses. Like many green politicians (Bookchin, for example), Leopold finds the Marxist critique of the capitalist position useful but would reject the statist alternative offered by socialism. And like our contemporary deep ecologists, he wants "something broader."

The exact identity of Leopold's "second group" is left vague, purposely vague since the degree of additional breadth that the group has achieved is "admittedly in a state of doubt and confusion." There is no doubt that an implicit appeal to scientific learning is present in Leopold's construction of this "second group" that "regards the land as a biota" and a "collective organism." Like Thomas Berry, but unlike many other deep ecologists who share his interest in nature mysticism, Leopold does not shy away from the use of scientific terms. A scientist himself, he has not lost his faith in science as a progressive force. He nevertheless expresses his need for something even broader, so he subtly builds into his argument a mystical element. Donning the prophet's mantle himself, he alludes to Ezekiel and Isaiah, both of whom figured the land of Israel as the collective body of the people, which could be alternately blessed or cursed by God (see particularly Ezekiel, chapters 20–22 and 34–35). Like Wendell Berry, another latter-day prophet, Leopold understands the land as the provider of identity, but he adds the references to science and thereby broadens the appeal of his work, at least potentially.

Moreover, Leopold entertains a covert line of reasoning that clearly distinguishes him from stricter traditionalists like Wendell Berry and marks him as a forefather of deep ecology. The argument parallels the historical case he makes for the extension of democratic ethics to the land. Just as the Old Testament ethic yields to the New Testament ethic which yields in turn to democracy which gives way finally to the land ethic, so Judaism was superseded by Christianity, which was replaced as a national mark of identity by the civil "religion" of democratic enlightenment but now should yield to "something broader"—perhaps a biotic religion or some form of reverence for life in all its forms. No doubt, opponents of the environmentalist movement—the Christian fundamentalist James Watt, for example—would be deeply upset by the suggestion that Leopold's argument reaches beyond ethics and touches theology itself, that ecology offers a *replacement* religion. As more than one Sierra Clubber has put it, "Conservation is a religious movement" (McPhee 208). Nevertheless, deep ecologists in recent years have not hesitated to expand upon the spiritual implications of their thought and to make full use of the rhetorical appeal to their readers' religious sensibilities. The scaffolding of the land ethic with natural mysticism that is merely hinted at in Leopold's essay becomes a full-blown connection between reverence and politics in deep ecology, with its "revival of Earth-bonding rituals, celebrating specific places" (Devall and Sessions 27).

Though his vagueness in calling forth "something broader" has provided a good deal of inspiration for deep ecology, the practical and scientific Aldo Leopold would likely have cringed at these developments. His rhetoric suggests that he already knew what the green politicians who have faced *Realpolitik* on its own ground have recently recognized: Appeals to either science or religion alienate some important portion of the politician's constituency.

Leopold's "Wilderness Experience": From Science to Mysticism

By stepping forthrightly into ethics from his position as a forester and in challenging his fellow foresters to do the same in articles in their professional journals (Potter 13–15), Leopold transgressed against the old boundaries of the scientific/engineering character and created a new field for scientific activism. As a model for other scientists, Leopold's eco-ethical system certainly has had some impact (see Potter, for example). But his influence among environmentalists is unsurpassed. Widely quoted at first, many of his views have now passed into the ordinary working vocabulary used in nearly every polemic devoted to environmental protection. No one has done more to shape the internal character of the movement. Stressing the impact of his importation of scientific understanding into the field of ethics, Linda Graber gives four major reasons for the popularity of Leopold's land ethic among environmentalists, particularly wilderness purists: (1) Purists are able, by applying this ethic, to see themselves as "the advance guard for a higher level of civilization, which is a more pleasant self-image than 'nature nut.' " (2) Leopold's notion of the wilderness as a baseline for land quality enjoys a nice conceptual fit with the idea of wilderness as ideal landscape or "sacred space." (3) Leopold's "fusion of the land ethic with the science of ecology lends the prestige of science to the purist's beliefs." (4) "Geopietistic mystic experience gains a code of moral directives based on scientific fact" (Graber 50).

Leopold's most direct treatment of wilderness is presented in his essay "The Wilderness Experience," one of the chapters in *A Sand County Almanac*. The essay begins with a dichotomy that ecospeak has rendered all too familiar—the division between the worker and the nature mystic, the core characters or archetypes of developmentalism and environmentalism, respectively. In a shrewd rhetorical move, however, a move that demonstrates his insight into the historical motives for seeking environmental amenities, Leopold treats the laborer and the nature lover not as two distinct characters that represent the opposing sides in a debate, but rather as two moods of one and the same character, a universalized human character. The portrayal thus leaves open the possibility for conversion of workers to the environmentalist political stance by suggesting to them that they do not have to give up their identity as workers to be lovers of nature. On the contrary, those who cherish the land most deeply are those who work with its resources directly and daily. Both work and contemplation are characteristically human activities, Leopole argues: "Wilderness is the raw material out of which man has hammered the artifact called civilization. . . . To the laborer in the sweat of his labor, the raw stuff on his anvil is an adversary to be conquered. So was wilderness an adversary to the pioneer. But to the laborer in repose, able for the moment to cast a philosophical eye on his world, the same raw stuff is something to be loved and cherished, because it gives definition and meaning to life" (264–65). Leopold uses this characterization as an ethical appeal, "a plea for the preservation of some tag-ends of wilderness, as museum pieces, for the edification of those who may one day wish to see, feel, or study the origins of their cultural inheritance" (264). When the developmentalists, one day hence, rest from work, they will mourn the lost wilderness if it is no longer there to fulfill their archetypal contemplative impulse.

In this passage and throughout *A Sand County Almanac*, Leopold's art lies in his suggestiveness, his technique of enfolding short but vivid hints of fuller reasoning within a highly condensed passage that communicates an overall feeling rather than an enumeration of particular points. First, there is the dichotomy of the laborer at work and in repose, a self-contained dichotomy that is highly attractive mostly because the great majority of readers will respond to it with nostalgic longing. We no longer work directly with the raw materials of nature. The division of labor in modern life is such that, even if we gather raw materials directly from the earth, someone else will process them; more likely, since we live as consumers of products that are far removed from their original raw state, only a powerful act of imagination will permit us to overcome our alienation from raw nature and imagine the original state of that which we use. In short, we neither stand at the anvil nor look with love upon unsmithed iron. Leopold's laborer is an anachronism.

But his very presentation of the laborer at work and at rest represents an effort to model the kind of imagination that would value the preservation of wilderness, an imagination that sees as a whole the process of converting natural materials to consumables. This holistic imagination perceives the human character as multi-faceted—both utilitarian and contemplative—and thus resists not only the undiluted *pioneering ethic*, which would deny the contemplative, but also the *mountaineering ethic* (of many Sierra Clubbers, for example), which would deny the pioneering impetus to conquer nature. Above all, the Leopoldian imagination would rebel against the *consumer ethic* that numbs us to the original value of both productive work and reflective contemplation.

There is also in the passage, however, an implicit privileging of the contemplative mood of the laborer over the productive mood, with the suggestion that the pioneering and consumer ethics have in America all but blinded the philosophical eye of the laborer. It is not the work of overwhelming nature as an adversary, but rather raw nature itself that "gives definition and meaning" to the life of Leopold's archetypal laborer. In a related passage, he writes, "Perhaps the most serious obstacle impeding the evolution of a land ethic is the fact that our educational and economic system is headed away from, rather than toward, an intense consciousness of land. Your true modern is separated from the land by many middlemen, and by innumerable physical gadgets. He has no vital relation to it" (261). Through his nostalgic vignette of the laborer, Leopold seeks to restore the visionary power of the consumer's blinded philosophical eye, much as the German Greens have made their political success by awakening their constituents' native affection for the Black Forest and other natural reflections of the German character threatened by such side effects of economic prosperity as acid rain. The preservation of remnants of the wild nature that "gives definition and meaning" to the laboring life is thus required if generations to come are to understand "their cultural inheritance." Without wilderness, our cultural artifacts and our cultural codes will seem baseless, cut loose from their original realism. Could the fairy tales of the Brothers Grimm have meaning to a child who has never known a forest?

Finally, there is a disarming, almost saintly modesty about the proposal put forth in the passage on the laborer's relationship to wilderness. Leopold calls for only "the preservation of some tag-ends of wilderness, as museum pieces." He is not out to deny the developmentalists their access to wide ranges of land. What industrialist could deny him his request for a few hundred thousand acres of wilderness space to be used as a natural museum, especially when, as he makes clear in the passages that follow, the preservation of many disappearing species of flora and fauna—including the wolf, the bear, the bison, and the long grasses of the prairies—would require so much more than he is asking for?

To the rhetorically suggestive worker-philosopher vignette that fixes the need for wilderness in human nature and culture and that makes its plea on that basis, Leopold appends three formal justifications for the preservation of wild places: *wilderness for recreation, wilderness for science*, and *wilderness for wildlife*.

The recreation argument builds directly on the concept of wilderness as a cultural need. "Physical combat for the means of subsistence was, for unnumbered centuries, an economic fact," he writes. "When it disappeared as such, a sound instinct led us to preserve it in the form of athletic sports and games" (269). Having thus appealed to the American love of sports, he extends the argument neatly: "Physical combat between man and beasts was, in like manner, an economic fact, now preserved as hunting and fishing for sport" (269). Wilderness areas are therefore important to the maintenance of the human character as "a means of perpetuating, in sport form, the more virile and primitive skills in pioneering travel and subsistence" (269). Then having appropriated for the environmentalist argument the human instinct for survival, to which the developmentalists have for so long laid exclusive claim, Leopold goes for the pathos of patriotism as well, suggesting that travel by canoe and pack train, two wilderness skills, are "as American as a hickory tree" (270). He declines to argue with those who would deny the importance of keeping these primitive arts alive: "Either you know it in your bones, or you are very, very old" (271).

He refuses, however, to leave a more crucial argument to be settled by the gut feelings of his readers and takes on directly "those who decry wilderness sports as 'undemocratic' because the recreational carrying capacity of a wilderness is small, as compared with a golf links or a tourist camp" (271–72). "The basic error in such argument," he asserts, "is that it applies the philosophy of production to what is intended to counteract mass-production." "The value of recreation is not a matter of ciphers. Recreation is valuable in proportion to the intensity of its experiences, and to the degree to which it differs from and contrasts with workaday life" (272). With this appeal to the human character—once again a universalization of middle-class nostalgia—Leopold has, in the space of a few short pages, brought his argument around to a direct confrontation of the developmentalist perspective.

Ironically, however, as he turns to the justification of wilderness for science, he appropriates the developmentalist drive for productivity and revives the old conservationist argument that our own health depends upon the health of the land. In a crucially important metaphor, he asserts that "the land is sick" (272). Before he developed this metaphor, Leopold must have carefully calculated the response of an audience that spends a fortune every day on health care. The success of medical science had never been more impressive than it was in the 1940s when Leopold wrote these essays. He ingeniously piggybacked upon that success by suggesting that wilderness provides the research lab for scientists engaged in curing the sick land, an "organism" whose own "health" is required for the health of the human organism. "The art of land doctoring is being practiced with vigor," he asserts, "but the science of land health is yet to be born" (274). "A science of land health needs, first of all, a base datum of normality, a picture of how healthy land maintains itself as an organism. We have two available norms. One is found where land physiology remains largely normal despite centuries of human occupation. I know of only one such place: northeastern Europe. It is not likely that we shall fail to study it. The other and more perfect norm is wilderness" (274). Leopold argues further that one wilderness is not enough since the scientist "cannot study the physiology of Montana in the Amazon; each biotic province needs its own wilderness for comparative studies of used and unused land" (274–75).

Risking anthropomorphism and the pathetic fallacy, Leopold's metaphorical attribution of "physiology" and "health" to the "organism" of the land is on one level an effective means of communication, using the familiar to explain the unfamiliar. On another level, it rhetorically forces an *identification* between human life and what is typically considered a dead, inanimate thing—the land—thereby undermining objectivism and asserting a radical subjectivism in the human relation to the land. Moreover, this rhetorical move is accomplished in a passage that uses science, the bastion of objectivism, as a justification for land preservation. By the time we come to Leopold's rationale of wilderness as a preserve for disappearing wildlife, we find none of his usual rhetorical gymnastics, but instead a simple statement of the need presented directly from the perspective of the environmentalist. It is likely that, at this point, he reckoned that he had either lost or totally captivated his audience.

In terms of the map we offered in figure 2 of our introduction (see page 14), Leopold's rhetoric involves the attempt to close the horseshoe into a circle, to complete the identification of science and deep ecology, to reconcile the view of nature as an object of study with the transcendental view of nature as spirit. Somewhat ironically, Leopold recognized that this identification and this reconciliation remain impossible so long as each of these traditions resists the subjectivism implicit in the humanist tradition, which has been accused of either giving rise to or acquiescing in the view of nature as resource. It is the human subject, the self, that provides the medium through which the objects of nature and the objects of the laboratory might be reconciled and mutually interpreted. The mind shapes the world and is shaped by the world. One of Leopold's deepest realizations is that scientific intelligence, like that of the worker-philosopher, simultaneously constructs nature and is itself re-formed by nature. The scientist's models—Leopold's own "biotic pyramid," for instance—are nourished by direct observations of nature and in turn provide categories for understanding those observations, for making them into new realities as scientific facts. Nevertheless, we cannot speak of observation and understanding without positing an observing and understanding *subject*, a mind, a self, the contemplation of which leads the scientist into history and ethics.

Leopold's rhetorical success lies in his expanding of the range of the modern self, whether that self is identified primarily as a scientist, an environmentalist, or a consumer. By positing an essential human unity outside of these narrow identities, Leopold seeks in Hegelian fashion a spiritual space where human beings can reach beyond their limits and ultimately come together. He identifies that sacred space with wilderness, which Graber has called "the Wholly Other." In the tradition of mysticism, the seeking of self leads to the loss of self in the contemplation of a higher entity. The potential danger of the drive for unity is that the obliteration of self is accompanied by the loss of a perspective from which to maintain the rules and codes of conduct that inevitably must attach to human behavior, even in mystical societies (Graber 11). The model of selfhood that Leopold has presented is thus developed as an interchange of activity and passivity, labor and reflection, that potentially engulfs, and validates, both the developmentalist and the environmentalist perspectives. In other words, a flow develops between self and other, a process of exchanging distance and identification. At each stage, the self is re-formed and renewed, as is the image of the natural world. By contrast, the agonistic rhetoric that pits environmentalist against developmentalist freezes the self in a role, a position in debate, and thereby closes off the possibility of reform and renewal.

In the work of Rachel Carson, we find a treatment of the objectivist personality more akin to that of standard reform environmentalism. Carson shows how scientific activity may be transformed into a scientific activism that will also replace the characteristic ethical and political passivity of research science. The danger in her approach is that, in her zeal to trade passivity for activism, she may well have underestimated the deep connection of passivity with reflection. Instead of synthesizing a new character, as in the transcendent logic of Leopold, she may have simply exchanged the contemplative character for an activist character. Whereas she moves closer to a firm grounding for political change—a ground elusive in Leopold's mystical wanderings—she loses breadth of appeal, though certainly not breadth of *effect*, for *Silent Spring* is perhaps the most widely read and most controversial of all books on environmental degradation.

Rhetoric and the Critique of Science in Carson's *Silent Spring*

Judging by commercial standards, Rachel Carson's *Silent Spring* is the most successful environmentalist book ever written. When it appeared in 1962, it found its way not only onto the best-seller list but also onto the desk and into the policy statements of President Kennedy (Brooks 285; Graham 51). It was not the first book to argue for the pervasiveness of the environmental danger; that distinction falls to Murray Bookchin's *Our Synthetic Environment*, published the year before Carson's book under the pseudonym Lewis Herber. But the rhetorical power of *Silent Spring* and the popularity of its author, who had distinguished herself as an excellent writer of science books for a general audience, played a crucial role in disseminating information about the threat of chemical toxins, thereby contributing in a direct and forceful way to growth of public environmentalism.

Like *A Sand County Almanac*, Carson's book established rhetorical conventions that would become standard fare in the environmentalist debate. Many of its rhetorical patterns endure to this day in activist writing and journalism alike. But whereas Leopold's work, despite its efforts at mass appeal, mainly influenced deep ecologists and other avant-garde thinkers on the fringes of the American public consciousness, *Silent Spring* captured the imagination of the mass culture, partly because of its dramatic, even sensationalist, language and partly because of its topic. The issue of the widespread dangers of toxins to public health and to nature in general clearly carried more weight with the mass of Americans than did Leopold's concerns over the elimination of the scientific, recreational, and cultural benefits associated with the preservation of wilderness. In this sense, *Silent Spring* was a culture-bearing book.

Carson's chief rhetorical strategy was the development of startling contrasts and dramatically rendered conflicts in a future-oriented report on environmental problems, a kind of apocalyptic narrative befitting a latter-day prophet. As science writing, the deep significance of *Silent Spring* lies less in its claims about the wrongdoings of the chemical industry, and more in its exposé of ideological diversity within the scientific community, its jeremiad on the dark side of technological progress, the claims for which had guaranteed the high reputation of science and engineering, at least after World War II. Hitherto, science writers had treated scientific researchers with awe and respect and had conventionally attributed to them much of the credit for societal progress in the West, thereby encouraging affluent audiences to identify with the overall project of science and technology. By dropping the signs of objectivist identity, Carson created a new image of science in the public mind and taught the scientific community that it was not a happy family and that it must confront its internal differences before it could hope to offer advice on national policy. In inverting the idea of progress, she led the liberal imagination toward the realization that its characteristic hope for the future may depend upon the development of environmentalist consciousness and action.

Carson's Apocalyptic Narrative

Silent Spring tells the story of how harmless, indeed helpful, plants and animals come to be destroyed by the careless use of strong toxins to get rid of insects and weeds that attack crops and that slow or disturb the development of roadways and housing tracts. It is not only the harmful effects upon wild nature that Carson is concerned about, however, but also the direct medical consequences for the very people who appear to flourish through the use of complex hydrocarbons for pesticides. The book is a frontal attack on developmentalists in industry and agriculture and the developmentalist mentality that has predominated among the liberal public. Carson suggests that, instead of living in fear of smallpox, cholera, and the plague, all of which have been overcome by modern medicine, "we are [now] concerned with a different hazard that lurks in our environment" (168). She quotes the medical authority, Dr. David Price of the United States Public Health Service: "We all live under the haunting fear that something may corrupt the environment to the point where man joins the dinosaurs as an obsolete form of life" (168). Far from relieving this fear or calming the audience, *Silent Spring* fosters anxiety with such assertions as "genetic deterioration through man-made agents is the menace of our time, 'the last and greater danger to our civilization' " (186).

The book's epigraph from Keats' "La Belle Dame Sans Merci"—"The sedge is wither'd from the lake / And no birds sing"—sets the stage for a rendering of the human tragedy connected traditionally with the desire to control nature through witchcraft (in the Faust legend, for example), the imagery of which Carson connects overtly with technological development. In this dark mood, the famous prologue, "A Fable for Tomorrow," unfolds. Though it is "for tomorrow," the prologue is written, like a foregone conclusion, in the past tense: "There once was a town in the heart of America where all life seemed to live in harmony with its surroundings" (13). A pastoral landscape of a farming community is laid briefly before our eyes. "Then," we are told, "a strange blight crept over the area and everything began to change. Some evil spell had settled on the community": the cattle and poultry died; the farmers and their families became ill with diseases that puzzled their doctors; the birds disappeared, leaving silence in the fields; the bees did not come to the apple blossoms; the fish disappeared from the streams; the roadsides, "once so attractive, were now lined with browned and withered vegetation as though swept by fire" (13–14). Who was responsible? "No witchcraft, no enemy action had silenced the rebirth of new life in this stricken world. The people had done it themselves" (14).

We might justify Carson's use of the past tense by suggesting her prologue's affinities with the genre of the fable, by saying it is a moral tale that harkens to a dead time that threatens to reemerge in the present. But Carson takes a more realistic tact in her own rationale for the tale, describing it as a kind of composite portrait. At the end of the prologue, she writes, "This town does not actually exist, but it might easily have a thousand counterparts in America or elsewhere in the world. I know of no community that has experienced all the misfortunes I describe. Yet every one of these disasters has actually happened somewhere, and many real communities have already suffered a substantial number of them. A grim specter has crept upon us almost unnoticed, and this imagined tragedy may easily become a stark reality we all shall know" (14–15).

"A Fable for Tomorrow" experiments with the shifting and overlaying of genres. In this brief introduction, she invites readers to adjust their perspectives and read what she has given them simultaneously as a fable, a psychologically realistic tragedy, and a report of actual occurrences. The effect of this kaleidoscoping of reading templates is to stir the emotions and to invite a variety of responses, building from vaguely felt discomfort and partial recognition to a deep intellectual and ethical concern over an actual state of affairs. Bringing a possible future alive in the present is the aim of this apocalyptic rhetoric. For an author who believes in the need for immediate action, present action that is required to forestall a human disaster in the future, this rhetorical tactic has great power; but as we shall see, it also involves some risks for writers who claim to base their conclusions in factual information. Facts do not exist in the future, only probabilities and projections. That is why, as Aristotle knew, deliberative discourse—that which debates the course of future action—always involves rhetorical appeals and can never be strictly descriptive and objective.[2]

The Science of Harmony vs. the Science of Control

Carson's prologue establishes dominant metaphors and themes that appear repeatedly, almost musically, as a structural feature of the book. The imagery of witchcraft and soul possession is picked up from the Keats poem and broadcast throughout the prologue: "Some evil spell had settled on the community" (13); "a grim specter has crept upon us almost unnoticed" (14–15).

Connected with the witchcraft theme is a dichotomy between natural harmony and the deadly unbalancing of natural forces, a contrast established early in *Silent Spring*. Like Leopold, Carson reminds readers of the human link with nature, the mystical union to be approached with care and devotion. She insists that "man, however much he may like to pretend to the contrary, is part of nature" (169). Unlike Leopold, however, Carson is just as interested in pointing out the enemies of nature as she is in winning converts to her cause. The enemies seek arrogantly to control nature but succeed only in turning nature against themselves and against the innocents whom they have duped: "The 'control of nature' is a phrase conceived in arrogance, born of the Neanderthal age of biology and philosophy, when it was supposed that nature exists for the convenience of man. . . . It is our alarming misfortune that so primitive a science has armed itself with the most modern and terrible weapons, and that in turning them against the insects it has also turned them against the earth" (261–62). Carson self-consciously undertakes her attack on the semantic level: "Can anyone believe it is possible to lay down such a barrage of poisons on the surface of the earth without making it unfit for all life? They should not be called 'insecticides,' but 'biocides' " (18). And she extends the fight to the syntactic level as well: "The bitter upland plains, the purple wastes of sage, the wild, swift antelope, and the grouse are then a natural system in perfect balance. Are? The verb must be changed . . ." (66). Neither before nor since Rachel Carson has environmentalism produced a writer with such a command of the language.

Her argument is directed against both the developmentalists proper—the "chemical salesmen and the eager contractors" (Carson 69)—and the scientists who collaborate with the pesticide industry, against whose objectivism she aims her lyrical irony. Not only do these collaborators practice bad science, backward "Neanderthal" science, she insists, but they also practice dishonest science: "Several American investigators conducting an experiment with DDT on volunteer subjects dismissed the complaint of headache and 'pain in every bone' as 'obviously of psychoneurotic origin' " (173).

Clearly indicating the direction of her rhetorical appeals, Carson ladles blame upon the scientist who sells chemical secrets for hard cash, while at the same time minimizing the guilt of the "average citizen" victimized by ignorance and overwhelmed by the rhetoric of the pesticide companies and developers. In this tactic, she is obviously aware of the centrality of the consumer to the historical development of environmentalism. Alluding to Vance Packard's early characterization of the consumer as victim in his popular book on the psychology of advertising, *The Hidden Persuaders*, she writes, "Lulled by the soft sell and the hidden persuader, the average citizen is seldom aware of the deadly materials with which he is surrounding himself . . ." (Carson 157). The scientists for hire cannot be so lightly excused. As if it were not enough to condemn them for collaborating with a "primitive," "Neanderthal" paradigm of science, Carson goes on to metaphorize their work as black magic, the ancient enemy of science, supposedly eradicated with the Enlightenment. Their chemicals are "brewed" like witch's broth (17) and sprayed across the earth as a "chemical death rain" (21), the polar opposite of Nature's nourishing rains; their primitive effects "bludgeon" the earth (64).

To the practitioners of this benighted science, the earth is already a dead thing. For the nature mystic, on the other hand, the earth is a body not unlike, and indeed intimately related to, the human body. The mystic desires the "health of the landscape" (Carson 69) and mourns the "scars of dead vegetation" (70), the "weeping appearance" of trees (71). Such mysticism, Carson seems to claim, can be reconciled with the scientific perspective, though not with the "primitive" science of control as practiced by applied researchers and engineers who have sold their secrets to the chemical industry. For they have ignored a new philosophy and theoretical outlook of basic science, in particular the new developments in holistic ecology.

Carson's apparent anthropomorphism may very well be justified under this shifting paradigm, as her colleague in holism, René Dubos, has suggested: "The phrase 'health of the environment' is not a literary convention. It has a real biological meaning, because the surface of the earth is truly a living organism" ("Limits of Adaptability" 27). Carson provides a summary and an explanation of her own ecological holism in this passage:

> For each of us, as with the robin in Michigan or the salmon in the Miramichi, this is a problem of ecology, of interrelationships, of interdependence. We poison the caddis flies in a stream and the salmon runs dwindle and die. We poison the gnats in a lake and the poison travels from link to link of the food chain and soon the birds of the lake margin become its victims. We spray our elms and the following springs are silent of robin song, not because we sprayed the robins directly but because the poison traveled, step by step, through the now familiar elm-earthworm-robin cycle. These are matters of record, observable, part of the visible world around us. They reflect the web of life—or death—that scientists know as ecology. (169–70)

Human responsibility to the earth, human participation in the cycle, is emphasized by the anaphoric rhythm of the passage: "We poison . . . We poison . . . We spray." Intentions may be good—no harm is meant for the robin—but cycles are set in motion that cannot be controlled. No esoteric scientific investigations are needed to show the results; they are empirically obvious. We can see for ourselves. We spray the elms; the robins die. The "we" of the passage—ubiquitous throughout *Silent Spring*—springs from a rhetorical tactic aimed to develop a seeing, understanding, reenlightened base of popular support, to "undupe" the "average citizen" and to undermine the authority of the hired expert, removing public support for the dead-earth school of science and developing a following for the live-earth school, a new, ecologically enlightened "we."

The ecologist's elaborate and organic view of the earth as structured by chains, cycles, and webs was news to many of Carson's readers in

1962. By relating this scientific perspective to romantic metaphysics, Carson discovered a familiar linguistic base and opened a fascinating set of rhetorical possibilities. The idea of ecology in the earth's body could be easily extended to the human body: "There is also an ecology of the world within our bodies" (Carson 170). The set of interrelations that determine whether we live or die—the metabolic cycles connected with the production of ATP in our bodies, for example, the "universal currency of energy" (185)—are profoundly disturbed by the intrusion of chemicals: "The fact of insecticide storage in the germ cells of any species should therefore disturb us, suggesting comparable effects in human beings" (186).

As a foil to the concept of controlling nature, Carson introduces the idea of "nature's control," the "relentlessly pressing force by which nature controls her own," mythopoeically summoning the familiar image of Mother Nature as a counterpoint to the witchcraft of the chemical industry. The word *tragic* appears over and over again, and it is connected with the apocalyptic tale of the arrogant brood of nature's human children turning against their mother, the sustainer of life. Pity and fear, said Aristotle, are the effects of tragedy, and these indeed are the responses Carson seeks in her rhetoric. We should simultaneously pity the earth and fear the results of her destruction. Our arrogance deceives us. We cannot live without her: "The shadow of sterility lies all over the bird studies and indeed lengthens to include all living things within its potential range" (Carson 101).

The Risks of Pathos

The appeal to emotion in the rhetoric of public debate is always risky. A writer who seeks one response may elicit a contrary one. Fear can cause readers to open their eyes wide or shut them tightly. Pity can draw them to a concern they have ignored or inspire defensiveness and lead them to deny their neglect and develop rationalizations, justifications, and excuses for their behavior. *Silent Spring* most certainly opened the eyes of many of the "average citizens" to whom it appealed. As Frank Graham has suggested in his documentary study of the development and aftermath of the book, "Rachel Carson uncovered the hiding places of facts that should have been disclosed to the public long before; she broke the information barrier" (xii). She left a mark so clear that "much of the subsequent history of pesticide policy is a response (pro and con) to Rachel Carson's judgment" (xii–xiii). But the book also shut some eyes. *Time* magazine—which would become the champion of the environmentalist cause in the late 1980s, but which represented a more conservative political constituency and agenda in the early 1960s—sharply criticized Carson's rhetoric in its "Science" section. The reviewer complained that "Miss Carson has taken up her pen in alarm and anger, putting literary skill second to the task of frightening and arousing readers" (qtd. in Graham 69). A reviewer in *Scientific American* dismissed the *Time* review as a proclamation of the "merits of pesticides" that failed to cite clear evidence of its claims, resulting in a "statement with which . . . no responsible scientist would want to associate himself" (qtd. in Graham 66). Yet the *Time* article implied that it was based on interviews with "responsible scientists": "Many scientists sympathize with Miss Carson's love of wildlife, and even with her mystical attachment to the balance of nature. But they fear that her emotional and inaccurate outburst in *Silent Spring* may do harm by alarming the nontechnical public, while doing no good for the things that she loves" (qtd. in Graham 69).

In fact, the scientific community was divided in their responses to the book. The very question of what constituted scientific responsibility was opened in a new and forceful way. Scientists engaged in pesticide research had always been portrayed as the helpers in the righteous cause of feeding the world. They were the architects of the "green revolution" (see Lester Brown, *Seeds of Change*). Now, according to Carson, these heroes had tragically slipped; the green revolution was proving to be a wrong turn, another show of human arrogance with uncalculated results that disturbed the balance of nature.

The public enemies identified by Carson had some eye-opening plans of their own. As Christopher Bosso has suggested in a 1987 study of pesticide politics, "Chemical industry and agribusiness executives viewed the problem largely in educational terms"; after *Silent Spring*, "the public needed to be reminded about the benefits of pesticides to society. . . . Their response was to flood the public consciousness with 'experts' who loudly refuted Carson's claims" (116). In addition to the counterattack of these experts for hire, however, many challenges came directly from the politically nonaligned scientific community in response to Carson's charge that scientists had been "trading their professional objectivity for obeisance to the needs of industry and of their own research-funding": "Carson's attack, in this sense, was not only on pesticides but on reputed 'scientific neutrality'; her thesis challenged scientific professionals to defend both their views and their integrity as scientists" (Bosso 116). The response was predictable; the book was criticized for its departure from the style and substance of the objectivist perspective, for its "mysticism" and "unscientific statements" (Graham 47).

Many of the scientific reviewers were particularly bothered by "A Fable for Tomorrow," the apocalyptic tale of a typical American town victimized by its own use of chemical pesticides. Roland C. Clement of the National Audubon Society remembered, "It just 'turned off' many scientists. The chapter is an allegory. But an allegory is not a prediction, which is what the literal-minded readers, with no background in literature, confused it with" (qtd. in Graham 64). Monsanto Chemical Company issued an extended parody of "A Fable for Tomorrow," which it called "The Desolate Year," its version of a tormented humankind living in a future without the benefits of agricultural chemicals and, in addition to publishing it in *Monsanto Magazine*, distributed over five thousand copies to newspapers and book reviewers across the country (Graham 65).

This propagandistic response may well serve to justify the fears of cautious scientific readers about the rhetoric of the apocalyptic fable. The effectiveness of the approach can hardly be doubted. As one legal educator, Ora Fred Harris, has commented, "In light of current uncertainties, perhaps the ideal teaching method to communicate the hazards of toxic waste exposure must entail the use of concrete, tangible, and highly dramatic environmental events that reflect the dire risk associated with toxic substance exposure" (102). Carson's composite story increased the drama of any of the single events she narrated in the book as a whole by concentrating them in the space of a few pages and in the setting of a single imaginary town. Her implication that the vignette could well be the future story of any number of towns increased the drama further by eliciting the fears and the insecurity that generally accompany the human contemplation of the future, the time of uncertainty in all matters but death. Since no facts exist in the future, however, no single writer can lay exclusive claim to an audience's imagination of future events, leaving open a vast field for rhetorical exploitation, irresponsible manipulation, and inherent inaccuracy. Even in the present, as Harris has suggested, scientific uncertainty broods over controversies involving exposure to toxic substances. "Science is not," Harris laments, "as exact as was once thought" (99). We cannot yet know what is a scientifically acceptable level of exposure. Thus the future is staked out as the battleground of the polemicists.

While a great number of scientific readers—many of whom, like Loren Eiseley and Julian Huxley, were also concerned with popularizing science—found no fault with the style or the content of *Silent Spring* (Graham 71), many others were deeply bothered by Carson's rhetoric. One of her own correspondents and sources of information, a Canadian zoologist, came out against the book after it was published, claiming that Carson had failed to adopt an editorial change he had recommended. He had wanted her to alter the title of the chapter "The Rumblings of an Avalanche" on the process of insect resistance to pesticides, a process he said was actually more analogous to the movement of a glacier. When he saw that the book retained the original wording, he complained in the Toronto newspapers that having been quoted in the book had hurt his reputation with his colleagues, and he concluded that Carson had been urged toward her hyperbole by "individuals concerned with wildlife work, having a vested interest in taking an extreme position" (qtd. in Graham 61).

Others questioned Carson's motives as yet the more sinister, implying that her campaign was part of a communist plot. Velsicol Chemical Corporation, the leading manufacturer of chlordane and heptachlor, two of the chlorinated hydrocarbons condemned by Carson and eventually banned for use in the United States, tried to keep Houghton Mifflin from publishing the book after its appearance in abbreviated form in three installments in *The New Yorker*. The Secretary and General Counsel of Velsicol wrote a letter to the publisher that concluded with this remarkable paragraph:

> Unfortunately, in addition to the sincere opinions of natural food faddists, Audubon groups and others, members of the chemical industry in this country and in Western Europe must deal with sinister influences, whose attacks on the chemical industry have a dual purpose: (1) to create the false impression that all business is grasping and immoral, and (2) to reduce the use of agricultural chemicals in this country and in the countries of western Europe, so that our supply of food will be reduced to east-curtain parity. Many innocent groups are financed and led into attacks on the chemical industry by these sinister parties. (qtd. in Graham 49)

In an effort to undermine Carson's ethos as a concerned, scientifically informed author, this passage associates her first with eccentrics in environmental fringe groups, "natural food faddists" and birdwatchers from the Audubon Society. Ironically, Carson herself was embarrassed by the gratuitous support she received from "food faddists, heath quacks and other cultists" (Graham 71). But more importantly, the Velsicol letter goes on to suggest a more distant and, given the context of the cold war, more damaging association: Perhaps not intentionally, but nevertheless certainly, books like *Silent Spring* are abetting the communist cause. The charge was serious enough and the political climate of the day heavy enough that Houghton Mifflin requested more specific information from Velsicol and had it reviewed by an independent toxicologist, who confirmed the accuracy of Carson's material. The plans for publication proceeded.

Means and Ends in Environmental Politics

As Bosso indicates, the response to *Silent Spring* and to the pesticide issue in general is a classic instance of what we have been calling the environmental dilemma:

> Problems like . . . the regulation of pesticides are both morally and technologically complex—what I call "intractable" problems. There is, in practical terms, no clear line here between means and ends, nor much agreement on either. . . . [It] is almost impossible to find agreement on ends because the ends themselves often are clearly incompatible: environmentalists may seek to rid the earth of pesticides at any cost; chemical firms may seek to maximize profits; farmers want inexpensive and effective pesticides to maintain high crop yields at lower costs; public health officials want to eradicate disease. Con-

> sumers, for their part, are caught in a bind; they want cheap food, which might lead them to support the wide use of pesticides, but they also fear the possibly carcinogenic effects of pesticide residues in that food or in the environment.
>
> Value conflict is accompanied by disputes over means and methods. Whose scientific data are more "correct"? Which analytical techniques do we accept as valid? Who decides? (xiii)

The agonistic rhetoric of the exposé, of which *Silent Spring* is a fine example, must ever rest on the assignment of praise and blame in an effort to influence decisions about public ends and means. It fosters controversy and divides perspectives, often attempting to arrange disparate interests into a clearly demarcated pair of opposed parties—environmentalist and developmentalist, for example—thus mobilizing citizens for a quick decision one way or the other, as is required in moments of crisis. Whereas academic commentators like Bosso (and us) try to delineate ethical viewpoints obscured by having been lumped together into a master perspective designed for political action in a situation perceived as a crisis, more recent contributors to the discourse on the environment have been trying once again to create a consensus, a "we," among groups formerly at odds with one another or with radically different interests. But the negative strain of the environmentalist movement that may be traced to *Silent Spring* also remains visible and active, a persistent "we" of public action, pointing fingers at "you" and "them."

In this process of building constituencies, as Bosso suggests, two kinds of appeals occur, one favoring arguments oriented toward ends, the other opting to focus on means. Ends-oriented arguments deal with the moral issues involved in public decisions; means-oriented arguments deal with the technical issues. To focus exclusively on moral issues is to drift toward stridency—the tendency of writers like Wendell Berry, who has said that "the basic cause of the energy crisis is not scarcity; it is moral ignorance and weakness of character" (13). To focus exclusively on technical issues, on the other hand, is to drift toward the kind of mechanistic reductiveness against which Berry writes in reaction, the kind exemplified in one science textbook's insistence that overpopulation "is a biological problem" with "biological solutions" (Norstog and Meyerriecks 620).

Carson's tendency to moralize the issues of pesticide management was likely part of her program to get people to adjust their thinking according to a new ecological paradigm, to see that to hurt the earth ultimately means to hurt ourselves. She was saying that we cannot treat the earth as a thing to which we can apply technical arguments only; our relation to the earth is a moral relation. The problem of reader response to this rhetorical position is that, unused to the new ecological paradigm, many members of Carson's audience must have perceived her moralization of natural relationships as strident and therefore suspiciously nontechnical.

Despite the slippage toward stridency in *Silent Spring*, however, Carson's technical objectives for social action were in fact quite clear and surprisingly modest. She essentially recommended three things: 1) extended scientific research, informed by the holistic paradigm of recent ecology; 2) more caution in testing chemicals before they are introduced into the environment; and 3) tougher standards for applying chemicals in agriculture and development, with tougher enforcement of those standards.

Of these recommendations, only the third would be considered controversial by practicing scientists and engineers, even those employed in industry. Beneath all three recommendations, though, there is an assumption that in many ways justifies the chemical companies' defensiveness about the book: the program of research, caution, and control favored by Carson is unlikely to be realized unless it is funded and implemented by strong, centralized government. Carson's interest in *publicizing* the debate therefore may well represent a covert or even unconscious interest in *socializing* research and control in the form of government regulation in matters where the public environment is concerned. In this respect, Carson's book is a clear spiritual forerunner of Garrett Hardin's influential essay, "The Tragedy of the Commons," first published in the journal *Science* in 1968. Hardin argues that we can no longer tolerate the frontier mentality that leaves the treatment of "the commons," our shared natural environment, to private conscience. Responsibility must become "the product of definite social arrangements," of mutually agreed-upon "coercions" (45).

By the late 1960s, this kind of statist argument, with its roots in the turn-of-the-century conservationist movement (Hays, *Conservation*), was again becoming an integral part of the environmentalist program for reform. Although Leopold and Carson, the scientifically informed predecessors of this late version of reform environmentalism, assumed the need for some public regulation, their rhetorical interest lay primarily in effecting attitudinal and philosophical changes in their readerships. In many ways, it was the rhetorical form of their writings, especially their lyrical romanticism and seeming subjectivity, that made them controversial as scientific writers. By the time we get to writers like Garrett Hardin and Barry Commoner, we find that the controversy has shifted to their politics. Impatient with attitudinal change, they call for radical social action.

From Science to Socialism: Barry Commoner's Environmentalist Writings

Like Rachel Carson, Barry Commoner began to question the place of science in American society early in the 1960s. His considerable skill as an expositor of difficult scientific concepts earned him a large readership. He became a regular contributor to *The New Yorker* and was frequently invited to participate on television panels and talk shows. By the 1970s, he had developed a full-blown critique of post–World War II social institutions—primarily the marriage of profit-dominated capitalism and science-supported industrial technology. These institutions had, in his estimation, created the environmental crisis. Delivering his critique with plain talk and scientific reasoning, with frequent appeals to common sense, Commoner has tried in every way to mitigate the ultimate drift of his thinking toward socialism as a solution to America's ecological problems. And despite his controversial political stand, perhaps even because of it, he has remained a popular figure in the mass media and has become the leading spokesman for science-based reform environmentalism.

Balancing the Concerns of Science and Political Rhetoric

In the early sixties—when the essays collected in his book *Science and Survival* first appeared—Commoner was adding his voice to Rachel Carson's in the effort to develop a scientific self-critique.

Citing the 1965 power failure that afflicted the entire northeastern U.S. and the appearance of thyroid problems in people who were exposed as children to fallout from the testing of atomic bombs in Nevada, Commoner brought public attention to "blunders which have begun to mar the accomplishments of modern science and technology." With the authoritative voice of a biology professor at a major university, he asked the question, "Is science getting out of hand?" and demanded that scientists respond to "the surprising failures of some of the new large-scale technological applications" by "applying the traditional skepticism of science to the recent growth of science itself" (*Science and Survival* 3, 6, 48).

For all his occasional rhetorical flare, however, Commoner has always carefully qualified his critique of science. First of all, he has shown not the least interest in abandoning science as a whole; on the contrary, he strongly affirms the place of responsible science in dealing with the complex problems of environmental and technological crises (*Science and Survival* 104). Indeed he insists that there is "a clear connection between our recent technological mistakes and the erosion of basic principles of scientific discourse" (63). He defends particularly the principle of free and communal inquiry, which has been undermined by the secrecy demanded in scientific work sponsored by industry and the military, and the commitment to basic research, which has been hurt by the privileging of applied research and engineering in the funding structure of American science. He concludes that "the capability of science to understand nature . . . is being damaged by the pressure of political goals" (106).

Yet, he argues, science cannot now afford to back out of its political and social involvement. In a passage that indicates the rationale for his own ethos, his own involvement in public life, Commoner writes:

> The impact of modern science on public affairs has generated a nearly paralyzing paradox. Despite their origin in scientific knowledge and technological achievements (and failures) the issues created by the advance of science can only be resolved by moral judgment and political choice. But those who in a democratic society have the duty to make these decisions—legislators, government officials, and citizens generally—are often unable to perceive the issues behind the enveloping cloud of science and technology. And if those who have

> the special knowledge to comprehend the issues—the scientists—arrogate to themselves a major voice in the decision, they are likely to aggravate the very threats to the integrity of science which have helped to generate the problems in the first place. (*Science and Survival* 108–9)

Commoner attempts to break the hold of this dilemma by asserting that the "scientist does have an urgent duty to *help* society solve the grave problems that have been created by the progress of science" and that this help should create an "informed citizenry" (*Science and Survival* 109). Scientists, then, should serve an advisory role as teachers of decision-makers: "Margaret Mead, the anthropologist, has called scientists' efforts to alert citizens to these issues and provide them with information needed for evaluating the benefits and hazards of modern technology 'a new social invention.' This may turn out to be the one invention of our technological age which can conserve the environment and preserve life on earth" (*Science and Survival* 120).

From his earliest efforts to inform the public of the hazards of technology, the applications of science that Commoner has attacked were the "new" and "large-scale" technologies that had grown to become an integral part of American life since World War II (*Science and Survival* 48). In his widely read book of 1971, *The Closing Circle*, he takes up this theme in what he calls an effort "to find out what the environmental crisis *means*" (11). Commoner offers a homey version of the mystic connection between organic and inorganic existence in his definition of the ecosphere as "the home that life has built for itself on the planet's outer surface" (11). In this metaphor and in his interpretation of the environmental crisis as "a sign that the finely sculptured fit between life and its surroundings has begun to corrode" (11), he implies the artistry of nature. The power of the scientist-critic lies in the ability to read the "signs" of nature. If on the other hand, the scientist-technologist lacks a view of the whole, the ecosphere comes to seem "a curiously foreign place"; life is seen as a series of separate and atomized events, each with a separate cause rather than as an interlocked web of causes and effects. Such flawed understanding implies the deadly alienation of humankind from the rest of creation, with human progess defined linearly, cutting through the finely netted web of life. Commoner writes, "Here is the great fault of the life of man in the ecosphere. We have broken out of the circle of life, converting its endless cycles into man-made, linear events: oil is taken from the ground, distilled into fuel, burned in an engine, converted thereby into noxious fumes, which are emitted into the air. At the end of the line is smog. Other man-made breaks in the ecosphere's cycles spew out toxic chemicals, sewage, heaps of rubbish—testimony to our power to tear the ecological fabric that has, for millions of years, sustained the planet's life" (12). Humankind's technological arts stand in destructive opposition to nature's artistry. Nature is the creator; human civilization is the destroyer.

Like Carson and Leopold, Commoner neatly integrates this science-based but romantically resonant vision of nature with his more practical critique of technological practices. But unlike Leopold, Commoner is not trying to forge a philosophical version of scientific ecology, by which to contribute to the development of an environmentalist perspective. And unlike Carson, he is not concerned with producing a piece of high literary art. His work is more politically practical and more definitely aimed at influencing actions.

He seems, however, a bit uncertain about how to proceed—a scientist out of his discourse element. In *The Closing Circle* (which, like so many important forays of scientific writers into the public press, was originally published in *The New Yorker*), passages on the philosophy of technology (such as the one just quoted) serve mainly as a rhetorical frame for the practical critique. The circle-of-life imagery is designed primarily, it would seem, as a device to enliven what might otherwise be a dull accounting of a statistical critique of human actions that degrade the environment. In one sense, the introduction to *The Closing Circle* is an effort to produce a journalistic "lead" as part of his intention to write in the genre that he or his editors have deemed most appropriate for generating information in the public forum—the feature story or magazine essay.

There is some internal evidence that Commoner was never comfortable with the rhetoric he felt was necessary to gain public attention for the important information he had to communicate about the environment. After introducing his aim to describe "some of the damage we have done to the ecosphere," he almost apologizes that "by now such horror stories are familiar, even tiresome" (*Closing*

Circle 13). "Much less clear," he says, "is what we need to learn from them, and so I have chosen less to shed tears for our past mistakes than to understand them" (13). He thus tries to distance himself from the pathos of writers like Rachel Carson and to maintain his stance as a scientist, a purveyor of logos. His aim, he says, is primarily "to understand" rather than to influence action, so it is no wonder that later in the book we find him hesitant to move into the role of social advisor.

He is particularly anxious about making assertions concerning the future. Consider in the following passage, for example, the number of qualifications and disclaimers he attaches to his prediction, as well as the self-distancing rhetoric implied in the use of the phrase "the environmentalist," the pejorative analogy involving the "occult seer," and the use of the pronoun "one" as the subject of the fourth sentence:

> These are the kinds of considerations that lead to my judgment that the present course of environmental degradation, if unchecked, threatens the survival of civilized man. Although it might be convenient if the environmentalist, like some occult seer predicting the end of the world, could set a date for this catastrophe, the exercise would be futile and in any case unnecessary. It would be futile because the uncertainties are far too great to support anything more than guesses. One can try to guess at the point of no return—the time at which major ecological degradation might become irreparable. In my own judgment, a reasonable estimate for industrialized areas of the world might be from twenty to fifty years, but it is only a guess. (*Closing Circle* 232)

The last sentence, with its tissue of qualifications—"in my own judgment," "might be," "only a guess"—appears to take back as much information as it gives.

A similar stance prevails in a discussion of possible damage to human fetuses due to a process by which one kind of plastics is fabricated. Commoner attempts to curb the full effect of the potentially apocalyptic rhetoric: "[The story] is *not* reported here in order to suggest that we are all about to perish from exposure to plastic automobile upholstery" (231).

In the passage that follows this sentence, he continues to qualify the effect of the story (in the phrases we have italicized) while at the same time pumping up the rhetoric with competing hyperbolic language (in the phrases we have placed in bold type):

> [The story] warns us that the **blind, ecologically mindless** progress of technology has **massively** altered our daily environment in ways that *may, much later,* emerge as a **threat to health.** Unwittingly, we have created for ourselves a **new and dangerous world.** We *would be wise* to move through it *as though* **our lives were at stake.** (231)

Such sentences appear to be the work of a reluctant advisor. The discourse of activist alarm in effect competes with the discourse of scientific caution.

Softening the Controversial Politics of The Closing Circle

But is this coyness itself a rhetorical guise? As we move through *The Closing Circle,* the rhetoric intensifies and the critique is focused ever more clearly. The chapter immediately following the introduction offers an excellent piece of teaching—a relatively straightforward, plainly delivered lecture on ecology, which, for the benefit of the lay reader, Commoner reduces to a series of familiarly worded and therefore memorable "laws of nature": (1) "Everything is connected to everything else" (33). (2) "Everything must go somewhere" (39). (3) "Nature knows best" (41). (4) "There is no such thing as a free lunch" (45). The laws are simple, commonsensical, perfectly clear to the scientifically untrained reader, especially to the careful observer of nature. The major authors in the American literary heritage of the last century intuited these laws, the set of relations we now call ecology. Commoner mentions in particular the insights of Walt Whitman, Henry David Thoreau, Herman Melville, and Mark Twain (*Closing Circle* 46–47). But the *effects* of these simple laws are so complex that even the best-trained and most experienced scientists seem baffled in dealing with them: "Few of us in the scientific community are well prepared to deal with this degree of complexity. We have been trained by modern science to think about events that are vastly more simple—how one particle bounces off another, or how molecule A reacts with molecule B" (21).

The lesson on the complexly webbed set of interrelations in the ecosystem, which seems politically innocuous, is followed immediately by a series of chapters on specific instances of environmental degradation and assaults on the web of life—the nuclear arms race, Los Angeles air, Illinois soil, Lake Erie water—in which seeds of political activism are planted and nurtured little by little. Like a typical scientific paper, which suspends conclusions until the data have been presented, the full force of the political argument is reserved for the last chapters. But each chapter offers a series of partial conclusions that build upon one another as we move toward the conclusion.

At the end of the chapter on Los Angeles air, for example, we are given two partial conclusions, or themes, that are developed fully in later chapters. The first theme concerns the relation of environmental politics to issues of social justice: "One thing that clearly emerges from nearly all statistical studies of the effects of air pollution on health is that they are most heavily borne by the poor, by children, by the aged and infirm; the most striking effects of air pollution on health seem to occur where the victim's health is already precarious. Certain features of social progress, such as improved nutrition, living conditions, and medical care, are known to improve human resistance to disease and thereby to improve health generally. In a sense, air pollution has a similar—but opposite effect on human health. It destroys social progress" (*Closing Circle* 79). By the time we are midway through the book, Commoner is ready to develop this theme further and to challenge the political commonplace that environmentalism is a "motherhood issue," an issue that no one can really oppose, that it is a diversion from other issues of social justice like the opposition to poverty and war (207). He asserts that, on the contrary, "as we begin to act on the environmental crisis, deeper issues emerge which reach to the core of our system of social justice and challenge basic political goals" (210–11). By the end of the book, he devotes an entire chapter to the "economic meaning of ecology," in which he cites one case after another of how perfectly good, relatively inexpensive products have been driven off the market by more expensive, technologically advanced products that offer no significant improvement in performance and are significantly more damaging to the environment: soap, for example, has been replaced by detergents; synthesized chemical fertilizers have replaced natural ones; large, high-powered automobiles have replaced more energy efficient ones (259–63). What has been improved by such replacements has been neither environmental conditions nor the advantage to the consumer. What has been improved is the profit margin of the manufacturers and distributors of the products.

The second theme initiated in the chapter on air pollution and developed cumulatively over the rest of the book is the cost, social and actual, of the new technologies, costs obscured by the hard sell of those who profit by technological change. Both the agents of the change and the economic/social nature of the effects are handled gingerly in this early chapter, however, which simply reports the paradoxical nature of technological progress: "Air pollution is not merely a nuisance and a threat to health. It is a reminder that our most celebrated technological achievements—the automobile, the jet plane, the power plant, industry in general, and indeed the modern city itself—are, in the environment, failures" (*Closing Circle* 80). One chapter after another reiterates this theme.

Then, again in the chapter on the relation of economics and ecology, Commoner draws together his themes on the two great enemies of environmentalism and the social justice movement in general—business and technology, or to pick up Commoner's alliterative thread—pollution, productivity, and the profit-seeking of private enterprise:

> The crucial link between pollution and profits appears to be modern technology, which is both the main source of recent increases in productivity—and therefore of profits—and of recent assaults on the environment. Driven by an inherent tendency to maximize profits, modern private enterprise has seized upon those massive technological innovations that promise to gratify this need. . . .
>
> The general proposition that emerges from these considerations is that environmental pollution is connected to the economics of the private enterprise system in two ways. First, pollution tends to become intensified by the displacement of older productive technologies by new, ecologically faulty, but more profitable technologies. Thus, in these cases, pollution is an unintended concomitant of the natural drive of the economic system to introduce new technologies that increase productivity. Second, the costs of environmental degradation are chiefly borne not by the producer, but by society as a whole. . . . A business enterprise that pollutes the environment is

> therefore being subsidized by society; to this extent, the enterprise, though free, is not wholly private. (*Closing Circle* 267–68)

Thus, "the emergence of a full-blown crisis in the *ecosystem* can be regarded, as well, as the signal of an emerging crisis in the *economic system*" (277; italics added).

Commoner as Liberal Environmentalist

The critique of technological capitalism leads Commoner toward a consideration of the theory and practice of political economy, broached superficially in the last two chapters of *The Closing Circle* and more fully, though also through an oblique (if not evasive) rhetoric in two later books, *The Poverty of Power* and *The Politics of Energy*. Commoner allows two points to the traditional defense of capitalism: (1) "The *practical* points of environmental pollution in industrialized socialist nations are not basically different from those typical of an industrialized private enterprise economy such as the United States"; "environmental pollution in the USSR," for example, "is following about the same course that it has taken in capitalist countries" (*Closing Circle* 278–79). (2) "Both socialist and capitalist economic theory have apparently developed without taking into account the limited capacity of biological capital represented by the ecosystem" (281). But he goes on to declare that "the *theory* of socialist economics does not appear to require that growth should continue indefinitely" (281), and he implies strongly that the socialist outlook is more compatible with the kind of steady-state economy needed to live in harmony with the ecosystem.[3]

Nevertheless, Commoner remains consistent within his role of scientific advisor, recommending neither particular economic reforms—though these, he feels, will be needed ultimately—nor even legislated environmental actions; he focuses instead on "the actions that such legislation is supposed to induce . . . , including: systems to return sewage and garbage directly to the soil; the replacement of many synthetic materials by natural ones; the reversal of the present trend to retire land from cultivation and to elevate the yield per acre by heavy fertilization; replacement of synthetic pesticides, as rapidly as possible, by biological ones;" and so on (*Closing Circle* 281–82).

Commoner surely recognizes, however, that his engineering recommendations, if implemented, would uproot the practices of capital-intensive, energy-intensive, technologically based business and industry. Like Rachel Carson, he suggests that these practices ignore the requirements of the natural ecosystem as developed in the new paradigm and as encapsulated in the common language of his four laws of ecology. Rather than developing his political critique by overtly negating the negative, by dismissing capital-intensive processes, he asserts a positive argument and implies the negative: "Harmony with the ecosystem may often be enhanced with the use of new processes which, while taking advantage of the best available scientific knowledge and technological skills, are relatively labor-intensive rather than demanding intensive use of capital equipment and power" (*Closing Circle* 290). Farmers, for example, will not have to go broke buying equipment to increase their yields so that the energy companies, equipment manufacturers, and food brokerage systems can profit by their loss.

Commoner concludes *The Closing Circle* with his most passionate and politically heated rhetoric: "The environmental crisis is somber evidence of an insidious fraud hidden in the vaunted productivity and wealth of modern, technology-based society. The wealth has been gained by rapid short-term exploitation of the environmental system, but it has blindly accumulated a debt to nature . . . a debt so large and so pervasive that in the next generation it may, if unpaid, wipe out most of the wealth it has gained us" (295).

The agents of "the insidious fraud" and "exploitation" who have "blindly accumulated a debt to nature" are never named, though it is implied that the guilt lies with all of "us." But in other parts of the book, victims and innocents are named who would logically be excluded from blame—the poor in the inner city who suffer undeservedly from pollution, the people in developing countries who have yet to know the affluence of life in technologically advanced societies, farmers who have resisted the claims of agribusiness and have wisely pursued organic techniques. The implied villains are the profit-seekers and their scientific and technological allies. They are the ones who must ultimately pay the "debt to nature." In a hyperbolic oversimplification, Commoner boldly engages the appeal of pathos in a technique we have encountered again and again in polemic speech and literature—the black-and-white, do-this-or-die presentation of "options": "Now that the bill for the environmental debt has been presented, our options have become reduced to two: either the rational, social organization of the use and distribution of the earth's resources, or a new barbarism" (*Closing Circle* 296; see Killingsworth, "Can an English Teacher").

When it comes time to account for the exact nature of this rational social reorganization, however, Commoner holds that "none of us—singly or sitting in committee—can possibly blueprint a specific 'plan' for resolving the environmental crisis" (*Closing Circle* 300). The "competence" needed to "change the course of history" is "reserved to history itself": "That we must act now is clear. The question which we face is how." The contingencies of ecospeak paralyze the author, freeze him in his role as contemplative scientist and create a set of overstated options that are are fearful to contemplate, much less to act upon.

The perspective developed in *The Closing Circle* is perhaps best classified as *liberal environmentalism*. A few traits may be isolated and criticized dialectically from other perspectives. The three articles of faith in liberal environmentalism, which we have already seen promoted in the reformist writings of special interest groups like the Sierra Club and which we will encounter again in science textbooks, include: (1) faith in nature as a self-regenerating provider of life, which, without human interference, would prosper as a steady-state system; (2) faith in science as the special knowledge of nature and the handmaiden of natural mysticism; and (3) dependence upon scientifically informed government regulation, as a necessary evil to counteract the negative influence of big money capitalism and destructive technology. On the first of these points, Commoner's faith is absolute: "Nature knows best" (*Closing Circle* 41). On the other two, he is somewhat ambivalent.

He certainly feels a deep uncertainty about the capabilities of science to influence social actions. Can scientists be trusted to apply the new paradigm of ecology and thereby correct what other scientists have caused by using knowledge to fuel the technological developments that have led to environmental degradation? Commoner attempts to solve this dilemma by dividing science into trustworthy practice and practice influenced by industry and military dollars—a division reflecting the standard opposition of basic and applied science. The scientist-for-hire has become a leading whipping boy of environmentalist critics, among whom Commoner must be counted. But on the other hand, everyone in America gets money from somebody, so that no one is utterly safe from this charge. Has the notoriety Commoner achieved as a scientific environmentalist, for example, affected his judgment in ecological matters? Does he not have a vested interest in using his authority as a scientist to put forward the environmentalist program of reform? "As *a biologist*," he writes, "I have reached this conclusion: we have come to a turning point in the human habitation of the earth. . . . [Continued] pollution of the earth, if unchecked, will eventually destroy the fitness of the planet as a place for human life" (*Science and Survival* 122; italics added). He nevertheless aims his critique at other scientists who have misused the very authority that he exerts. An alternative perspective, that of deep ecology, for example, might argue that modern science has had its day and has failed. Folk wisdom, the wisdom of poetry, mysticism, and the organic tradition of medicine and farming, might serve with renewed force as an guide for ecological reform.[4] Commoner himself suggests as much when he notes that the four "laws" of the ecosystem are as much the property of transcendental literature, notably that of Whitman, as they are the product of scientific theory. But Commoner's faith in scientific reasoning—typical of the environmentalist perspective and correlative with his own personal interest—ultimately remains firm.

His ambivalence emerges again on the subject of government regulation. Commoner teeters on the brink of socialism but resists in his early writings going over to the radical cause because of a liberal mistrust of coercion, whether it be instituted in corporate capitalism or in big government. This stance is most clearly presented in his now infamous denial that overpopulation is a major cause of environmental degradation, a position that he maintains to the present day (see his *Making Peace with the Planet*). In *The Closing Circle*, he argues that "population growth in the United States has only a minor influence on the intensification of environmental pollution. If United States agricultural and industrial operations were ecologically sound, the country could support many more people than it does now with far less environmental impact" (233). He finds the scientific literature on the population problem equivocal, so he opts for a political—specifically a liberal—rather than a scientific solution. Defending the right of the family to self-governance, he denounces programs that call for regulation of family size as the products of "political repression," asserting that Hardin's argument in "Tragedy of the Commons" for "mutual coercion, mutually agreed upon" is in fact a plan by which "the majority would need to be coerced by the minority" (*Closing Circle* 214). That Hardin's argument is arrived at by rational, scientifically acceptable procedures has no influence on Commoner, who insists on making his stand on morals and politics, upholding the classic liberal defense of the family's and the individual's sovereignty.

The position on population has drawn the fire of a number of Commoner's colleagues in science-based environmentalism. The rebuttal offered by the Stanford biologist Paul Ehrlich is perhaps the most damaging. Ehrlich, the author of the best-selling book *The Population Bomb*, suggests that Commoner dismisses the population question in order to sharpen his thesis that the blame for the environmental crisis should be laid at the feet of profit-hungry, technologically based, energy- and capital-intensive big business. Ehrlich notes two "unfounded assumptions" in Commoner's argument. The first is that, if people had stayed on the course they had charted before World War II, environmental impact would have remained small despite growths in population. Against this position, Ehrlich cites instances of human-caused ecological disasters spreading across history from the extinction of mammals in the early Pleistocene to the dust bowl of the 1930s. The second assumption is that postwar technologies have developed independently of the need to meet the demands of the increasing population. In fact, many of the attempts to increase productivity in agriculture may have, in Ehrlich's reading (as well as in the reading of Lester Brown in *Seeds of Change*), been forced by the pressures of the rising demand of ever larger world populations (Ehrlich and Holdren 86–87).

The argument over population lays bare the political foundation of Commoner's writings and threatens to reveal the seams of his critique of technological capitalism. His single-minded pursuit of the profit-seekers in American culture may indeed be viewed as a departure from his insistence on the complexity that characterizes the scientific paradigm of ecology and that, according to the argument of Ehrlich and other scientists, is essential if we are to cope technically as well as morally with the "storm of crisis problems" involved in the environmental crisis as a whole: "Complacency concerning any component of these problems—sociological, technological, economic, ecological—is unjustified and counterproductive. It is time to admit that there are no monolithic solutions to the problems we face. Indeed, population control, the redirection of technology, the transition from open to closed resource cycles, the equitable distribution of opportunity and the ingredients of prosperity must *all* be accomplished if there is to be a future worth having. Failure in *any* of these areas will surely sabotage the entire enterprise" (Ehrlich and Holdren 87). The line of reasoning exhibited in the last sentence is a better application of one of Commoner's "laws"—"Everything is connected to everything else" (*Closing Circle* 33)—than his own insistence on privileging the moral and political over the technological in a discourse that borders on stridency. Moreover, his liberal arguments contradict another of his laws—"there is no such thing as a free lunch" (45)—in the suggestion that we can have larger populations without suffering grave environmental consequences.

Beyond Liberalism

In his later writings, Commoner has sought to go beyond his early liberalism and to develop a full-blown environmentalist socialism. *The Poverty of Power* (1976), for example, a treatment of the ecological/economic aspects of the energy crisis of the 1970s, exhibits a rhetorical structure similar to that of *The Closing Circle* but progresses toward a distinctly more radical conclusion. It begins with an introduction that reticently avoids stating the book's controversial thesis but uses a rhetorical lead to interest the audience in the problems implicit in the energy crisis. Next appears an extended exposition in layperson's terms of thermodynamic theory, followed by an application of the general laws of thermodynamics to each of the major energy alternatives—oil, coal, solar, and so on. The book drifts into the economics of energy, moving ever more rapidly toward the negative conclusion that he could not bring himself to form in *The Closing Circle*. The chapter on economics concludes with a strongly political paragraph based on an analogy that appeals mightily to the scientific-technological perspective:

> When engineers want to test the strength of a mechanical system, they stress it until it breaks and thereby reveals where it is weakest. The energy crisis is such an "engineering test" of the economic system. The stress it has imposed on that system is the threatened shortage of energy—the inevitable result of our short-sighted dependence on non-renewable and technically unreliable sources of energy, and our grossly inefficient ways of using it. Modern production technology has transmuted the stress into a shortage of capital and jobs. This is an ominous metamorphosis, for it signifies that the economic system is unable to regenerate the essential resource—capital—which is crucial to its continued operation, or to serve the people in whose name it was created. What is now threatened is the economic system itself. This may be the true price of power. (*Poverty of Power* 234)

The appeal to the technological consciousness and to "the people" converges in the last chapter in the conclusion that "the energy crisis and the web of inter-related problems confront us with the need to explore the possibility of creating a production system that is consciously intended to serve social needs and that judges the value of its products by their use, and an economic system that is committed to these purposes. At least in principle, such a system is socialism" (258). Commoner's suggestion that "it may be time to view the faults of the U.S. capitalist economic system from the vantage point of the socialist alternative—to debate the relative merits of capitalism and socialism" is a far cry from a manifesto or a call for socialist revolution, but it is at least tacitly more radical than the calm rhetoric in this last chapter would suggest.

The seeds of radicalism have, by this time, already been nurtured in the earlier chapters. We have heard of "a fault that lies deep in modern society" (*Poverty of Power* 3); of "energy problems [that] will not be solved by technological sleight-of-hand, clever tax schemes, or patchwork legislation" (4), the usual remedies of the capitalist system; of "the failure of private enterprise" (119); of "a lopsided partnership between the private and the public sectors" (120). We have heard that the oil companies, like "a poorly trained birddog," are no longer "a reliable vehicle for the production of U.S. oil, for they seem to be interested less in producing oil than in producing profits" (62–63); and that "the giant corporations have made a colony out of rural America" (172). This is the ultimate attack on developmentalism, an effort to show that its foundations in technological capitalism no longer support its ability to provide jobs and money, that its strength in economics is no longer a valid rationale for its failure to provide a means of preserving resources and caring for the environment.

In short, Commoner urges the reader to make a choice between a steady-state socialism that is at least theoretically possible and a capitalism that has no hope of survival in a world of shrinking resources and increasing damage to the environment. For all his insistence about the urgency of the current crisis, the conclusion is rather weakly stated. The tentativeness of this socialist conclusion, an indirectness supported by the suspended thesis that is the rhetorical trademark of both *The Poverty of Power* and *The Closing Circle*, suggests either the author's own tentativeness or his fear of a nonreceptive audience and his worry about his role as a scientific advisor. Commoner's stance also resembles that of a professor wary of taking his students too quickly beyond their most cherished values in their initiation into new knowledge and new communal values.

Liberal Environmentalism: A Dominant Perspective in Science Textbooks

Despite the controversial political stance that Commoner has adopted, a stance that he himself often seems to reluctant to take, he and other architects of science-based social reform appear to have found a willing audience in the field of science education. It has become quite common for secondary school and undergraduate science textbooks to align ecology with social activism, thus substituting an engaged political perspective for the more traditional "apolitical politics" of scientific autonomy and objectivity. In light of this trend, Theodore Roszak's comment that ecology is a "radical deviation from traditional science" would seem less naive than it appears on the surface (Evernden 5). Explicit or implicit in the introductory biology and ecology textbooks we have surveyed is the argument that, because nature has been seriously disturbed by the life-style encouraged in technologically developed societies, citizens must cultivate scientific understanding and must support government intervention to protect the environment from ecologically unsound projects.

Informed Citizenry and Government Protection Against Developmentalism

The textbooks that we examined teach that natural systems are capable of organic adaptation and regeneration unless human activity interferes on a large scale with natural processes. The authors of one high school biology textbook take a grim view even of the effects that common human activities have had upon natural systems: "Humans have cut forests, cleared underbrush, and burned fields. . . . many bird habitats have been destroyed. People have drained marshes and lowered the water in ponds. . . . water birds can no longer find a source of food and nesting sites. In the past many thousands of birds have been slaughtered . . . conservation efforts came too late for some species" (Otto and Towle 768–69). The authors' sharpest criticism, however, is reserved for technological/industrial development:

> Pollution problems . . . have destroyed wildlife and made water unfit for use in many areas. . . . Oil slicks . . . foul the beaches and kill many organisms. (772)
>
> Most industrial wastes are not biodegradable. Power plants, steel mills, paper mills, refineries, and automotive factories are examples of industries that have used rivers for waste disposal. . . . pollution of streams with heavy metals . . . makes bathing in the water and drinking it dangerous. . . . Fish from the water are not safe to eat. (772)

The authors' observations ultimately lead them to the alarming conclusion that our society in its current form is on the brink of suicide: "Will the human race decline," they ask, "like the paramecia culture, poisoned by its own wastes?" (755–56)

As a corrective measure, the textbook implies its support for the liberal proposal that industrial development and other forms of human activity should be controlled by government intervention guided by scientific expertise: "It is impossible to eliminate air pollution entirely from an industrial nation such as ours [they write]. However, it must be reduced. Industry is being made to comply with laws . . . regarding almost all phases of air pollution. . . . Scientists are still studying this problem to see how the pollution of our ecosystem can be further reduced" (Otto and Towle 779).

Other authors state their advocacy of government regulation even more forcefully. Consider this statement from an advanced high school biology text, which adopts many of Commoner's arguments for the need to develop a "well-informed public" and for statism as a political system that "at least in theory" creates the conditions for a more balanced human relation with nature: "Although incentives for individual action are effective ways of getting most things done, environmental problems can be solved only by effective actions by the government, an agency that, at least in theory, can look beyond the immediate interests of individuals and plan for the long-term welfare of society. . . . In a democratic society, this can only be done with the support of a well-informed public that understands the biology of organisms and the intricate web that connects us to all living things" (Arms and Camp 878).

A textbook for junior high students takes up the same themes, focusing on the need to preserve wildlife and to protect the public right to environmental amenities: "In areas where the government and the citizens together have made a conscious effort to clean up the environment, positive results can be seen. . . . We must all be willing to invest time, effort, and money to pay the cost of cleaning up our environment. This cost may be high, but the cost of not cleaning up our environment is high also. It includes a higher death rate and higher medical costs to treat illness. It also includes the destruction of wildlife and the loss of a pleasant environment to live in" (Teter et al. 167).

The Image of Science

In the textbooks, science is uniformly portrayed not as the insulated research community that it has become in its specialized disciplines but rather as a reliable guide for personal and communal action; it is thus imputed to occupy a special advisory role in the social structure. The junior high textbook just quoted offers this observation:

> Careful study is needed whenever people want to do something that might upset the balance in a natural community. . . . it seems that most of the time these results do not begin to be seen until after a long period of time. In fact, upsetting the balance of nature is a process that may take place over a period of time as long as several hundreds of years. An example of this is the using up of our natural resources. . . . the fact that these changes may not be noticed for a long period of time adds to the problem. The longer it takes for people to realize that their actions are upsetting the natural community, the more difficult it may be to correct the situation. (Teter et al. 133)

The implication is that the "careful study" that science undertakes represents a better way for an unenlightened community characterized by blind action.

The grandchild of the Enlightenment and the Progressive Era, this theme appears again and again in the textbooks. Referring to population control and improvement of the quality of life in "underdeveloped nations," the authors of an advanced high school biology book assert, with no small degree of arrogance, "Whether biological principles will be applied in time to salvage the world ecosystem remains to be seen" (Norstog and Meyerriecks 620). Another advanced high school text makes the same point somewhat more modestly, concluding a discussion of the environmental and social results of population pressure with this statement: "Food and the income to buy food, are unevenly distributed between the rich and the poor. The problem is, therefore, more economic and political than biological. We can learn something of the solutions that will or will not work, however, by examining the biology of the situation" (Arms and Camp 873). A bold summary of this view of science as a privileged advisory institution is given in a college textbook for elementary science teachers: "Teachers have the responsibility to show students that scientific discoveries can also help to prevent the destruction of the environment. No scientist can know how valuable or dangerous his or her discovery will be in advance. But all citizens should insist that scientific discoveries be used to improve the earth's environment and produce a better life-style" (Carin and Sund 15).

Reasons for the Textbooks' Ethical and Political Engagement

The liberal environmentalist perspective emerges clearly as a system of values in the textbooks reviewed in our survey and represented briefly here. It is certainly not the only set of values set forth, but it is perhaps the most frequently stated and the most systematically reproduced. Why must textbook writers feel compelled to encourage the development of such values in what is generally regarded as a value-free discipline of study, the science of ecology? A few tentative conclusions are possible.

Any effort to make subject matter relevant to an audience involves a parallel attempt to create a foundation of shared values. This is one of the deepest lessons of modern rhetoric, especially as taught in the work of Kenneth Burke. The textbook writers may reason that, if students are going to get interested in science, it must be related to their own interests as whole human beings—social and political beings as well as scientific investigators. Scientists who write and read articles for *Ecological Monographs*—after years of training in a culture that depends upon a strict division of labor—may have learned to shed their roles as parents and citizens in order to participate in the worldwide community of ecological scientists. But young students must be courted rhetorically to ensure the continuation of scientific research.

Such an approach would be consistent with a general shift from objectivism to values-oriented education. Specialists concerned with environmental education overtly subscribe to the position that "the ultimate goal of our educational systems must be a citizen that lives all facets of life in a manner that is humanly successful yet ecologically sound" (Roth 4; see also Reynolds and Wootton 10). One research team defines environmental education as "a process whose purpose is the creation of a citizenry who can help resolve environmental issues" (Monroe and Kaplan 38). A prominent educational theorist, Parker Palmer, has suggested that community-conscious education should replace the epistemological "objectivism" that is derived from science but that, according to Palmer, pervades all areas of the standard curriculum. This objectivism involves distancing the knower from the world for the purpose of preventing knowledge from becoming prejudiced and biased; analyzing information; and experimenting with the information, moving its parts around, thereby reshaping the world into a more pleasing place. Palmer asserts that this mode of knowing breeds competitive individualism and a loss of integrated vision, which preclude the existence of any real community of learners. He implies that objectivism prevents students from learning the types of community values necessary for civic virtue. As an alternative, Palmer suggests that new curricula should encourage a new mode of thought—a type of cooperative learning that incorporates intimacy—a way of personally implicating oneself with the subject.

Two problems emerge if we attempt to interpret the textbooks we examined as exemplars of the new human values approach to education: (1) In the general biology textbooks—and in science education as a whole, so far as we can determine—not all topics are treated within a social context. Human reproduction, for example, is rarely connected, at least not overtly, with birth control techniques or with the values of planned parenthood or with a discussion of abortion. A special emphasis on values is reserved for environmental science. (2) The textbooks do not encourage consensus formation or the right to disagree or an understanding of a range of perspectives—the kind of democratic education that Parker Palmer has in mind. Instead, they unvaryingly advocate a single political outlook. The values of liberal statist environmentalism are privileged over all others.

As a relatively new field, environmental science is still in its formative stages. The inevitable unsettledness of the discipline appears in the textbooks as an uncertainty about the boundaries of the science's conclusions in the overall academic division of labor. Will environmental studies finally develop as basic science, an applied science, a form of engineering, a social science, or an interdisciplinary pursuit that refuses to be confined within one field and that thus is among several recent academic courses of study (including, for example, discourse theory and semiotics) that radically question the old division of labor? The latter development has already taken shape in several college programs and in such college textbooks as *The Environment: Issues and Choices for Society* (ReVelle and ReVelle 1988), which combines a rudimentary education in ecology with a study of human activities relating to the environment.

Uncertainty urges us toward familiar ground. Textbook writers, knowing the importance of environmental studies but uncertain of their curricular status, have grounded their work in the political and ethical categories of ecospeak. They have submitted to the tendency to divide the environmental debate into two camps, the environmentalist and the developmentalist. They have forsaken objectivism not so much because of a desire to replace it with a values-centered or humanist epistemology, but because it has no real status in public debate over the environmental dilemma. Both environmentalists and developmentalists in the mass media and the arena of public politics *appeal* to science, but neither side really accepts or even understands the traditional scientific outlook on the world. The character of this debate has found its way into science textbooks and may yet go even further in undermining the well-guarded objectivity of basic and applied research in scientific ecology.

One other possible explanation for the pervasive influence of liberal environmentalism in the textbooks should be considered. Could it be that the scientific community in fact shares this set of values, but that in their specialized work, scientists *assume* the values rather than argue them outright in the manner of the textbook authors? According to this interpretation, the textbooks form a system of indoctrination into the scientific community, while the discourse of research maintains conventions of objectivity without subscribing to objectivist values.

Against this hypothesis, our research suggests that environmentalist values have very little influence on mainstream scientific research and that all such ethical systems are thoroughly repressed in late undergraduate and graduate education, during which time the conventions of the research discourse are learned. Although like most academics, ecological researchers in the universities tend toward liberalism in their personal politics, they subscribe in their professional work to a strict code of ethics that strives to insulate them from the influences of public rhetoric. The very structure of the research discourse, like the discourses of many aboriginal communities, is armored against intrusions from forces outside the community. In chapter 3, after we have given a more detailed account of the structure of this discourse, we will return to the relationship of the community of scientific researchers to the discourses of science teaching and environmentalist activism. Then in chapters 4 and 5, we will show how the separation of the internal rhetoric of science from the public rhetoric of environmental politics stands in the way of the liberal goal of creating an open democracy of informed citizens.

Scientific: The principal motive is the establishment and organization of knowledge. To the extent that it attempts to influence human action and opinion, or relies for verification upon communal values . . . , it moves in the direction of rhetoric.

—Walter Beale,
A *Pragmatic Theory of Rhetoric* (96)

If the technical expert, as such, is assigned the task of perfecting new powers of chemical, bacteriological, or atomic destruction, his morality as technical expert requires only that he apply himself to his task as effectively as possible. The question of what the new force might mean, as released into a social texture emotionally and intellectually unfit to control it, or as surrendered to men whose specialty is professionally killing—well, that is simply "none of his business," as specialist, however great may be his misgivings as father of a family, or a citizen of his nation and of the world.

—Kenneth Burke,
Rhetoric of Motives (30)

3

Scientific Ecology and the Rhetoric of Distance

Facts and Rhetoric in Scientific Discourse

It was once common to think of scientific discourse as somehow above rhetoric—the practice of the courtroom, the political arena, the pulpit and revival meeting—or at the very least, different from rhetoric. In recent years, however, the view has emerged that science definitely has a rhetoric but that it is *rhetorically distinct* from public rhetoric and from writing that *appeals* to science without *doing* science. Writing that appeals to science and the writing of science are produced and consumed from perspectives that are all but mutually exclusive.

In this chapter, after briefly elaborating some points of recent theory on scientific discourse, we will consider the work of several ecological scientists for whom writing is a means of doing science. Their work is generally characterized by a set of preferences that creates a distance between the activities of mainstream science and those of the public realm of environmental politics by favoring theoretical arguments over naturalistic description, by using a specialized technical language instead of employing familiar uses of the same or similar lexicons, by ignoring general human interest in a science of natural (that is, nonhuman) relations, by privileging basic over applied research, and by making a strong distinction between refereed and nonrefereed literature. With an understanding of this special discourse in the background, we can proceed in chapters 4 and 5 to a study of discourses that describe or appeal to science without doing science—namely, science journalism and the instrumental documents of government control.

The Rhetoric of Scientific Writing

How does scientific writing differ from other kinds of writing? Until recently, rhetorical theorists have been satisfied with the view of scientific discourse that distinguishes it on the basis of its superior objectivity or "referentiality" (see Kinneavy). This view accepts an understanding of reality as given, "out there," something to be described. If the activity of science is primarily careful observation, then the discourse of science is a careful recording of that careful observation. In this view, the goals of scientific rhetoric—if science can be said to have a rhetoric—would thus be clarity and thoroughness. Moreover, scientific writing could be grouped together with other forms of reporting—journalism or police reporting, for instance, or any kind of writing whose primary concern is "the facts," whose principal aim is to *refer* to a world rather than to *construct* a world.

The main problem with this understanding is that, while it gives us a way of distinguishing scientific writing from political speeches, poetry, and fiction, it does not allow us to make a strong distinction between writing that is taken as authoritative by scientists themselves and writing that presents scientific information for the purpose of influencing social and political actions (the discourse of scientific activism, for example, or environmental impact statements produced by government bureaus, or fact sheets submitted by oil companies to the local newspaper). Nor does this category of "referential discourse" allow us to distinguish between scientific writing and writing that describes scientific or factual information for a nonscientific audience (science textbooks or science journalism, for example). No working scientist would accept a theory that failed to make such distinctions.

The first step in showing that scientific discourse-practice is different from that of simple reporting is to allow that scientific writing is rhetorical, even though to make such a claim is to break with a tradition of discourse taxonomy that goes back at least as far as Aristotle. Nevertheless, it is clear that scientists engage in rhetorical discourse as defined by Walter Beale: "The kind of discourse whose primary aim is to influence the understanding and conduct of human affairs. It operates typically in matters of action that involve the well-being and destiny of communities (and of individuals within them);

and in matters of value and understanding which involve the communal or competing values of communities" (94). In this sense, *all* discourse is to some degree rhetorical since writing and speaking are themselves "human affairs" and forms of conduct. All discourse is *performative*, a means to an end. The difference between science and other discourses is a difference in the kind of practices and communities it supports and moves forward. It is read and written from a different perspective with a different agenda for action and a different set of values. The communal destiny and communal values with which scientific rhetoric is concerned are the destiny and values of the scientific community itself. So the question becomes, how is scientific activity distinct from political or journalistic activity?

Science as a Distinct Agenda for Action

There has been, in the last fifty years, much dispute among philosophers of science about the exact nature of scientific activity. Karl Popper has argued that its chief characteristic is falsification, so that when one scientist puts forth a hypothesis or theory, other scientists attempt with all their resources and vigor to prove it wrong. In a famous demurral from Popper's influential view, Thomas Kuhn substituted a normative image of scientific activity less flattering to the individual working scientist than the Popperian image of heroic struggle but more in line with the idea of science as a social activity. That is, following a revolutionary breakthrough in theory—the work of a Copernicus or a Darwin—"normal science" proceeds to interpret the world in light of the new theory and does so until emerging facts place this normative theory or "paradigm" in crisis and a new paradigm comes to the forefront. Irme Lakatos preserves elements of both the Popperian and the Kuhnian schemes in his concept of the rational progress of science depending on the success or failure of research programs. Accordingly, the value of a given program and its durability is determined by its predictive power; a research program fails and is abandoned when its predictions are found again and again to be wrong.

What all of these theories have in common is a view of science as *containing its own history*. All suggest that science is more or less insulated from extrascientific influences.[1] For writing to be scientific, it must do science. That is, it must perpetuate a research program and have no interest in influencing actions that lie outside of that research program—no interest in the kinds of social and ethical actions that form the object of environmental impact statements, no interest in the entertainment value of science news, no interest in the aesthetic or poetic value of the reality it portrays. An assertion that the ultimate aim of science is to contemplate or describe "reality" or "nature" becomes a metaphor that is useful to the scientist as a means of distinguishing scientific activity from other activities. It is a means of value labeling. Activities valued by the community are granted currency when this label of favor is attached. To use the categories established by Walter Beale, we can argue that research is considered "scientific" or "objective" if it is considered to be free from *instrumental* action (in engineering technology and government, for example), the goal of which is to control the state of things; from *policy* action (in politics, for example), the goal of which is to change the state of things; and from *poetic* action (in novels, for example), the goal of which is to supplement the state of things. In fact, science may still do all of these things, but only within the limited realm of action certified by the scientific community. Its characteristic discourse may well involve *instrumental* efforts to control the state of things (in experiments that manipulate nature); *policy* efforts to change the state of things (in arguments among members of the scientific community); and *poetic* efforts to supplement the state of things (in predictive models and theoretical speculations about events not yet observed or recorded). The *content* of these modes of action will differ drastically from the content of political action or poetic action in the extrascientific world. That content is controlled carefully by the scientific community. "Objective" is the compliment paid to the work of a researcher who is "one of us, doing what we want to do in the way we want to do it."

How Scientific Rhetoric Creates the Facts

In a particularly insightful analysis of the social construction of scientific objectivity, Bruno Latour suggests that, when it comes to the "facts," scientists are realists. But the facts, far from being the unchangeable solid realities of the positivist, are rather the conclusions of arguments that have been settled within the scientific community. "Cold" science is the retrospective contemplation of certified truths, of "black boxes" that are "closed." The popular view of science, the notion that science deals with settled facts, is therefore at least half right. It is also half wrong, for it ignores "warm" science, or science in the process of fact making. This portion of scientific activity involves the reopening of factual black boxes, the introduction of new experimental data, the forming of new arguments, the reconstructing of reality. In the controversies of "warm" science, the participants throw over their realism and become relativists. Scientific reality, according to Latour, is "what *resists*" (93). Arguments clinched long ago resist reopening, while arguments in the process of settlement are vulnerable and nonresisting (see also Fleck 98; Bazerman 61–62).

"As long as controversies are rife, Nature is never used [by scientists] as the final arbiter since no one knows what she is and says," Latour claims. "But once the controversy is settled, Nature is the ultimate referee" (97). Science differs from politics and poetry because it appeals to "nature" or "reality" only after a certified set of procedures and a certain number of investigators and a certain quality of argument have produced a certified version of nature or reality. Moreover, it admits into its arena of contemplation no constructions of reality developed in competing traditions and perspectives. No creationist, for example, driven by hypotheses suggested in humanist and poetic texts, will ever be admitted into the field of scientists engaged in constructing an image of the world's origins. Creationism is a theological, not a scientific theory.

What is considered to be factual in scientific research is not necessarily the same as what is considered factual in other discourses produced from other perspectives. In making this observation, we are beginning to establish the ultimate paradox of the scientists' construction of their own discourse community: It is founded upon a politics that demands freedom from politics and a value system that claims to be value free.

Scientific Discourse: A Case

To add concreteness to our description of the rhetorical process of fact building and discourse certification in the scientific community, we present in this section a profile of the activities and an analysis of the rhetoric of seven ecological scientists at Memphis State University. We developed these profiles by reading the scientists' recently published refereed papers and then interviewing them in order to clarify their overt aims and intentions in writing these papers and to discover their general objectives in research. What emerges, we hope, is a cumulative portrait of the characteristic discourses connected with ordinary scientific activity at a fairly typical state-funded university.

From our analysis of the papers and more especially from the interviews, we established a set of themes that appeared frequently as major concerns of the scientists and that had a strong bearing on our study. The themes tended to develop as a set of oppositions. We will focus on five opposing pairs:

- Natural history vs. theoretical science
- Familiar language vs. scientific language
- Human interest vs. natural science
- Applied research vs. basic research
- Gray literature vs. refereed literature

The comments of the scientists we interviewed suggested that the values of the general community of ecological scientists tend to be strongly associated with the terms on the right in these pairs (theoretical science, scientific language, natural science, basic research, refereed literature).

Later we will show that many journalists, politicians, and environmental activists, in their quest for a global solution to the environmental dilemma, fail to recognize the strength of these oppositions in the minds of research scientists and thereby earn the contempt of the scientific community. Science may well represent a strong example of communicative action, as Habermas suggests (*Theory of Communicative Action* 2.91–92), but part of the strength and success of this discourse community derives from a scrupulous patrolling of its borders against intervention by extrascientific interlopers.

Natural History vs. Experimental Science

Horkheimer and Adorno have written, "In science there is no specific representation. . . . Representation is exchanged for the fungible—universal interchangability" (10). In a developing modern science like biological ecology, the move is certainly in this direction: out of the field—the buzzing and blooming confusion of specific and varied details (to paraphrase William James)—and into the laboratory, the world of controlled data and descriptions of models that apply not only to a pasture just east of Memphis, Tennessee, but to any plot of land with similar characteristics.

Raymond D. Semlitsch, one of the scientists we interviewed, studies frogs, salamanders, and most recently insects. But he is not interested, as his nineteenth-century counterpart would have been, in cataloging and naming species and placing jars of collected specimens in the glass cabinets of museums. He is interested in "complex life cycles," in the description of organisms that have at least two stages of life and that live in two different habitats (amphibians, for example, that live part of their lives in water, part on land). He can range in his research from salamanders to mosquitos because he is a "process- and concept-oriented" theorist rather than a specialist "interested in a specific problem or a specific species." Thus he is an experimental scientist concerned above all with theorizing about evolutionary patterns.

In his article "Fish Predation in Size-Structured Populations of Treefrog Tadpoles," Semlitsch describes an experiment on the ability of large tadpoles to survive in the presence of small-mouthed fish (bluegills) as compared to the survivability of small tadpoles. The experiment could not have been done as a simple observation of fish and tadpoles in a lake. Under such conditions, no reliable counts could have been made. And even more important than this practical difficulty, there would have been a methodological objection to simple observation: If data were collected in the uncontrolled setting of a single pond, this information would be pertinent only to that specific pond. As Semlitsch says, "I'm not studying a specific system or a lake or pond, but common processes in all lakes or ponds." Any finding that he records should be "important enough that it will occur in any lake or pond, artificial or natural." So he and his colleague, J. W. Gibbons, ran a double experiment, Gibbons making his observations in outdoor artificial ponds into which a given number of fish and tadpoles were placed, and Semlitsch running similar tests in plastic dishpans in his lab.

The results are summarized in tabular form in the published paper. Centered on the page in the "window" of text that the eye seeks first (centered vertically, and just above horizontal center), and framed by an ever-so-slight addition of white space, a wide column of numbers in a large table presents the data that establish forcefully the relationship of tadpole size to survivability—a string of zeroes in the column indicating how many small tadpoles (averaging about twenty-seven milligrams) survived in the presence of large fish, yielding as the reader's eye moves down the column to an ever-higher rate of survivability among medium-sized and large tadpoles. With great economy and visual power, the table shows the main result of the experiment, the finding that bigger tadpoles have a better chance of surviving in the presence of gape-limited predators and of succeeding to their adult phase.

The rhetorical tightness and cleanness of the tabular presentation of data are reinforced by the use of closely parallel headings and subheadings in the text and by the parallel structure of emphatic interpretive sentences describing the significance of the major findings. The significance is reported in the first sentence under each heading: "The body size of tadpoles had a dramatic effect on their survival in the presence of fish" (Semlitsch and Gibbons 322). "The size of fish predators significantly affected ($P = 0.0019$) the survival of tadpoles in the laboratory experiment . . ." (323). "The full sib family from which tadpoles came did not significantly affect ($P = 0.1208$) their survival when exposed to fish predators . . ." (324). The findings are important, the conclusion tells us, because they "indicate that environmental factors affecting the growth rate of tadpoles can dramatically alter their vulnerability" (325) and may therefore "act as an agent for natural selection to increase growth rate" (326).

When we arrive at this conclusion, we begin to grasp the deep purpose of counting tadpoles in plastic dishpans in a biology lab. The scientists' aim is to advance evolutionary theory. Semlitsch is involved in the effort of ecologists to demonstrate how environment affects evolution. Far from being rhetorically neutral, the presentation of his data is vigorous and challenging, especially his use, and even repetition, of the term "dramatic" and his emphatic structuring of his tables and the results section of his text. He is clearly engaged in argument, not in mere description. He is especially interested in staking out a territory in evolutionary theory, in asserting the presence of environmental factors in natural selection—hence the importance of the negative result involving full sib families.

This conflict between ecologists and geneticists—a conflict that, since its origin in the arguments of Lamarckians against orthodox Darwinians, has taken various turns and forms in the history of biology—is taken up even more vigorously in the work of William H. N. Gutzke, who like Semlitsch, specializes in herpetology and environmenal science. The agonistic rhetoric of the debate is well exemplified in this passage from his paper, "Influence of the Hydric and Thermal Environments on Eggs and Hatchlings of Painted Turtles":

> Since the "Great Synthesis," evolutionary biologists have been primarily concerned with genes and the mechanisms by which their frequencies are affected by natural selection. Environmental effects have been relegated to mere causes of "noise" in the system. However, data are now appearing which indicate that genetic differences are not the sole important agent for intraspecific variability. Because natural selection can only act on the phenotype of an organism, the capacity for phenotypic variation in response to environmental variation (phenotypic plasticity) is of great evolutionary significance and must be considered if the complex interactions that enable organisms to survive in their environment are to be understood. (402–3)

Though many of the stylistic features we will encounter in our description of the environmental impact statement in chapter 5 are present in this scientific paper—the favoring of passive constructions and the obliteration of human subjects in sentences, for example—none of the blank neutrality of tone appears. In reading, we are aware of an engaged author, devoted to asserting a particular perspective. The paper, rather typically, begins with an introduction that places previous research in question (in a "review of the literature") and finishes with a conclusion that asserts forcefully a revision in the overall research program of evolutionary scientists.

The commitment to a point or a perspective and the consequential rhetorical heat of these papers are related to the highly theoretical slant of this work. Though rhetoric has been traditionally associated with the muddy worlds of politics and ethics, where "opinion" predominates over "fact," argument in science is taken up with passion on topics that have at one time or another been considered matters of fact. As Latour has suggested, the topics that are argued in theoretical science were once "closed"—the dominance of genetics in natural selection, for example—but have been reopened by a new research program, in this case, the program of environmental science. When we asked Gutzke's opinion regarding Latour's claim that every word and every figure in a scientific paper is devoted to the clinching of an argument, he agreed without hesitation.

The public perception of science as a field of settled facts results from a confusion of the type of science we have been describing with what is generally referred to as *natural history*, the detailed account of a specific corner of the world in terms of the settled "facts" about that region. In our interview with Semlitsch, however, he cautioned against oversimplifying the distinction between natural history and theoretical science. Using an approach that had obvious affinities to our own, he suggested that we think of the distinction as a relationship along a continuum. Figure 3 represents the continuum we sketched out in our discussion.

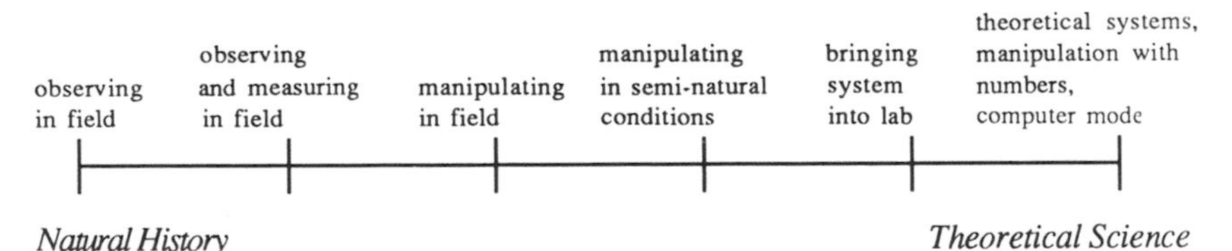

Figure 3. Continuum from Natural History to Theoretical Science

As scientists determine their activities along this continuum of practice, they must realize that they are involved in a system of trade-offs.

One set of certainties is sacrificed for another. At the point where you move out of the field and into "seminatural conditions" (like Gibbons' artificial ponds), "you are sacrificing realism for experimental rigor," as Semlitsch sees it, "realism for the ability to control parameters."

From the perspective of rhetorical analysis, this means that, as scientists move from natural history activities to those of theoretical science, they are leaving the realm of "cold science," the world of positivist fact, and entering the world of "warm science," of theoretical disputation. As scientists move into this favored realm, however, they distance themselves from the political and social disputation over the environmental dilemma, which occurs in the world of natural (and social) history. This distancing, one of the major factors in the modern construction of objectivity, casts doubt on the immediate usefulness of the conclusions and assertions of "warm science" in decision making about environmental questions. This is the first trap set for those who would make polemics out of the conclusions of a developing science. There are others.

Familiar Language vs. Scientific Language

People commonly think of the language of scientific literature as a jumble of latinate technical terms connected with inscrutable formulas. In the science of ecology, an uninitiated reader will be surprised by the relatively high percentage of recognizable terms. But this aspect of familiarity is deceptive; it is another trap for the unwary humanist or social ecologist.

In describing the theoretical aims of his work with complex life cycles, for example, Semlitsch speaks of the organism's "decision to go from one stage to another." The use of the term *decision* would likely offend the ecological economist E. F. Schumacher, who has criticized his fellow economists for similarly applying terms like "planning" to "matters outside the planner's control" (211). Certainly a tadpole cannot "decide" when to become an adult frog—at least not in the sense of the term as we use it to describe conscious human volition. But in another sense, in the scientifically certified sense, the *body* of the tadpole does make a kind of decision. There are probably more technical ways of describing what occurs, but they would be long and awkward. The familiar term becomes a shorthand expression for a series of events which scientific readers will grasp immediately but which lay readers are likely to puzzle over, dwelling upon the absurdity of the image of the tadpole's big decision to become an adult.

We discussed this anthropomorphizing effect of scientific language with Ronald Mumme, an ecologist who studies how birds' behaviors are affected by environmental and genetic factors. We suggested that, even in Darwin's own writings, the problem surfaces in the construction of the metaphor *natural selection*. An untrained reader of *The Origin of Species* is likely to perceive in Darwin's narrative both a theological residue, a substitution of Nature for God as the agent of creation, and a humanistic residue, a substitution of nature for the human breeder as the agent of selection in the famous comparison of natural and artificial selection in cattle breeding. Darwin and the scientists writing in his tradition frequently use what Mumme calls "a teleological kind of language" for "nonteleological" concepts; that is, they *write* as if nature were a purposeful force, working toward a *telos*, a preestablished end (like the ultimate good of all beings); but in fact, they *believe* that occurrences in nature are accidental, chance combinations of events that produce some good fits between an organism and its environment—combinations which lead to the perpetuation of the species—and some bad fits, which lead to extinction. *Natural selection* itself is a teleological term suggesting that nature has a plan, a favored outcome, but in Darwinian theory the term describes an accidental outcome (see Dawkins).

In using such terms, the scientist is counting on the audience's ability to read theory into ordinary language. "I'm writing in a convention," says Mumme, "using a verbal convenience for people used to this language." Not that technical language for these processes is unavailable. Any of the familiar terms could be "unpacked" and rendered in "unloaded descriptive terms," but the resulting prose would be "tedious and almost unreadable." Following are three typical passages from Mumme's articles on bird behavior. The "loaded terms" in these cases are those with resonances which, for the nonspecialist reader, suggest parallels to human behavior. For the lay reader, the effects of these terms are startling, misleading, and intrusive, but such effects are presumably absent for the scientific reader:

> 1. These findings are consistent with predictions of inclusive fitness theory, and have led to the suggestion that kin should receive favoured treatment in the form of not only cooperation but also reduced aggression or competition. However, selfish, competitive and manipulative behaviours can evolve among close relatives if the gain in direct (individual) fitness exceeds the loss in indirect fitness. (Mumme, Koenig, and Pitelka, "Reproductive Competition" 583)
>
> 2. The establishment and maintenance of territories requires significant investments of time and energy. . . . Because this cost of territoriality may reduce the amount of time individuals have to devote to foraging and breeding, a variety of ecological factors may influence the amount of time individuals devote to territory maintenance. . . . We attempted to evaluate the effects of three ecological factors, food availability, intruder pressure, and temperature, on the amount of time male Carolina Wrens (*Thryolthorus ludovicianus*) devote to vocal advertisement and defense of their territories during the winter. (Strain and Mumme 11)
>
> 3. In many animal species, males frequently use tactics to increase their confidence of paternity in the offspring produced by their mates. . . . Although apparently widespread, paternity assurance behaviours should be especially prevalent among species with male parental care. . . . In these species, males that do not take steps to ensure their paternity risk becoming victims of kleptogamy (cuckoldry), expending their parental investment on the offspring of the other males. . . . (Mumme, Koenig, and Pitelka, "Mate Guarding" 1094)

Apparently, talk of "selfishness," "costs," "kin," "investments," "advertisement," and "confidence" in bird behavior raises no objections and is transparently clear to the scientific readers of these passages. Even for technically trained readers, however, the picture muddies somewhat with the introduction of a term like *cuckoldry*, which must be coupled with the technical term *kleptogamy* in the last sentence quoted. Mumme tells us that the use of this term has been criticized. He also says that the effort of some scientists to substitute the term *rape* for *forced extrapair copulations* was a linguistic move that proved altogether unacceptable. Even for scientific readers, a term like *rape* is too radically invested with emotion to be looked upon with the dry detachment considered to be typical of the objective scientist.

This last example suggests that between the scientific usage and general usage of words, there is no absolute opposition but once again a continuum. At one end of this scale of acceptable diction, we would place purely technical terms, like *kleptogamy* (though even this term resonates with familiarity because of its etymological and semantic similarity to *kleptomania*, a term used widely in a culture concerned with private property). Other terms—*selfishness* and *confidence*, perhaps—would fall toward the middle of the scale, having no doubt strong personal and human connotations for scientists who use them technically, but not so strong as to interfere in the context of a scientific paper with their semantic propriety within a theoretical framework. Finally, on the far extreme of the continuum, are terms like *rape*, whose overtones of the conditions of everyday human life are so vivid and strong as to prohibit their effectiveness in functioning as quasi-technical terms. A term like *cuckoldry* would likely join *rape* at the extreme of prohibition were it as widely used in contemporary culture as it was in Shakespeare's day.

The continuum of usage hints at the opposition between behavioral study as a human science on the one hand and behavioral study as an animal science on the other. If certain language is partitioned for exclusive use in the human realm—thus reinforcing the dichotomy between human life and nature—so also are certain topics of human interest banned in "hard" behavioral science. This observation leads us to our next pair of opposites. Since the aim of theoretical science is to expand the generality of conclusions beyond the realm of single species and single environments, we might wonder, Why would behaviorists in the hard sciences create linguistic and experimental limits that prohibit entry into the human realm?

Human Interest vs. Natural Science

We took up this troubling question most directly in our interview with Gutzke. After the publication of several of his papers on the relation of environmental factors in reptile behavior and sexual development, he received "some very strange calls from around the nation," asking him for opinions on the causes of everything from mass murder to homosexuality. But like most hard scientists, his policy is to "stay away from social issues." We asked what constituted a "social issue." It is, he said, a topic that "deals with emotions and norms" of human behavior. "Scientific issues deal with data," he explained—data that can be looked upon as "mere" data, with the scientist following "an accepted statistical analysis," one that is "accepted by other scientists, experts in the field." However, since the social sciences deal with data, are they not also truly "scientific"? In principle, Gutzke said, the human sciences could be truly scientific, but in practice, there are problems. Above all, "your own social history is likely to bias your interpretation of data." He offered abortion as an example. It is an issue that "elicits strong emotions in humans," but abortions in the organisms studied in hard science are simply matters of death and life, uncolored by human feelings. In dealing with the sexual habits of adult bull snakes, you "don't have to deal with perceptions," Gutzke claims. But the phone calls did come in. Thus certain issues in behavioral ecology have inevitable resonances in the world outside the laboratory.

Consider another example. Beverly Collins, a plant ecologist we interviewed, conducted a series of experiments to determine how "canopy gaps, formed by the fall or snap of one or more trees, create patches of altered environment within forests" (Collins and Pickett 3). The experiments involved cutting trees to create openings in the forest ceiling so that additional sunlight and rainfall could be admitted to the underlying layer of vegetation. This work has clear significance for those interested in defending or attacking current practices in the cutting of timber. But this application does not interest Collins, who claims to be "apolitical" and who told us, "Unless we can understand the basic processes for how environments respond, we cannot even *understand* the applied problems."

Somewhat ironically, her work entails manipulative practices that distantly parallel the actions of many of the villains of the environmentalism, the North American lumber industry and the South American cattle industry. "Sometimes," Collins explains, "the best way to get at the questions is to do manipulative field research." Her manipulations of natural environments—meager though they are—have not escaped association with the exploits of clear-cutting. In an otherwise objective review of recent work in the field, an author in the British journal *Nature,* writing of Collins' work, let slip a phrase widely used to describe large-scale cutting of trees—"forest destruction"—then passed blithely back into the standard reviewer's style of scientific reporting: "In a piece of *experimental forest destruction,* Collins and Pickett created gaps in hardwood forest in Pennsylvania by cutting canopy and understory trees at 1 metre above the ground. Herb layer vegetation shows very little response to such gaps. It is reasonable to suppose that soil disturbance is needed if there is to be additional recruitment from the seed bank" (Moore 313; italics added). Whether the phrase in question was intended as a joke or an implied criticism, or whether it was simply an unconscious slip, neither we nor Collins can tell. The association of scientific activity with the questionable practices of industry is, however, clear enough to be disturbing for a scientist interested in keeping clear of human interests and ethical questions in research about plant ecology.

If this association is troubling for a scientist primarily engaged in "basic research," what about those engaged in what is commonly called "applied research" in ecology?

Applied Research vs. Basic Research

This opposition formed the most widely mentioned and emotionally loaded topic in our discussions with the scientists. Semlitsch, Gutzke, and Collins all made a definite point to distinguish themselves as basic researchers. Mumme did the same, explaining that a recently funded study on the habitats of the Florida Scrub Jay, an endangered species, is his first excursion into applied research. But three other scientists we interviewed—Stephen Klaine, Mark Hinman, and Gary Wein—are forthrightly engaged in the kind of applied research that government officials and reporters in the mass media seem to consider the major business of scientific investigation. Our interviews showed, however, that these applied scientists were either somewhat defensive about the status of their work in the scientific community or were openly critical of the "elitism" of basic researchers. In mainstream science, applied research is seen as a marginal activity. It slides away from the theoretical aim of purely scientific discourse and engages the question, *What should people do?* It thereby undermines conventional scientific attitudes about the strong dissonances between the opposing pairs we have been sketching, sometimes favoring natural history over theoretical science, familiar usage over scientific diction, and, most importantly, human interest over natural science.

Stephen Klaine and Mark Hinman are environmental toxicologists and coauthors of several recent papers about a project designed to monitor pesticide migration in west Tennessee. It is one of the largest such projects in the world and could have an important effect on agricultural practices in a region dominated by farming—the Mississippi Delta. The following passages from the introductions of two of the papers show a clear connection between the research and the problems that we associate with the environmental dilemma. In the first passage, the discussion is even framed according to the classic opposition between high productivity and environmental degradation:

> Pesticide use has resulted in significantly higher crop yields as agriculture increasingly relies on the chemical industry. Pesticide use increased 40-fold from 1946 through 1976 and this trend has continued to the present. . . . The dependence of agriculture on synthetic fertilizers and pesticides will continue to increase as greater productivity is demanded per unit of land. Concurrent with this increase will be a greater flux of chemicals from treated land to untreated land, surface water and ground water.
>
> Characterization of agricultural pesticide migration has become a primary concern. . . . (Klaine et al., "Characterization of Agricultural Nonpoint Pollution: Pesticide Migration" 609).

> Nonpoint source pollution can be defined as the diffuse input of pollutants that occurs in addition to inputs from undeveloped land of similar genesis. . . . agricultural nonpoint sources of pollution significantly altered water quality in 68% of the drainage basins in the United States, and in nearly 90% of the drainage basins in the north central region of the United States. . . . agricultural sources are probably the major contributors of suspended and dissolved solids, nitrogen, phosphorus and associated biochemical oxygen demand loadings in U.S. waters. . . .
>
> West Tennessee has the highest erosion rates in the country. . . . This high erosion rate and the related agricultural chemical burden

are the primary reasons why west Tennessee rivers and streams have the worst water quality in the state. . . . (Klaine et al., "Characterization of Agricultural Nonpoint Pollution: Nutrient Loss" 601)

Immediately following the introductions are sections on the methods and instruments used to measure contents of the runoff from an experimental field. Results and discussion are given in a single additional section, breaking with the conventional practice in reports on basic research of separating results and discussion or conclusions into different sections. The focus is clearly on results, which largely substantiate the literature but also hint that the overall picture of nonpoint source pollution is worse than previously depicted. The papers try above all to build a more complete data base of information pertinent to the topic.

Thus, instead of a theoretical conclusion to these papers, we get descriptive passages with little argumentation. As an alternative to advancing theory, we might expect the authors to offer a set of recommendations for action, but they stop short of this. To offer recommendations would mean that they had moved all the way out of the field of science and into engineering. In fact, there are engineering professors among the coauthors. Though the human interest of their work has increased, however, it has not overwhelmed the interest in natural science. The point at which they stop might well be described as *natural history*. In the terms developed along the continuum of figure 3, this work falls just to the right of absolute natural history—"observing and measuring in the field"—with a minimum of manipulation involved in setting up an experimental watershed.

Klaine, the senior scientist on the project, has spent his career developing methods and collecting and analyzing data about the effects of human-generated substances on biological systems, especially plant systems. He has pioneered several bioassays to determine how algal cultures are affected by contact with such human products as wastes from coal gasification facilities. We asked him if his work was oriented more toward the solving of human problems than was the work of the basic researchers we had interviewed. His answer was equivocal. "You have to understand the problem before you can solve it," he said, and this understanding constitutes the aim of applied research and also sets the limits of objectivity to some degree. How will the data generated by the pesticide migration project be used? *Eventually* it will lead to "better methods for avoiding migration." But the work of applied science does not go that far. Nor can reliable recommendations yet be made.

Klaine affirms that science has an advisory capacity in the overall society in that the data he produces will be used to alter current methods of agriculture. "One would hope that a person making a regulatory decision would have the scientific data to base that decision on." It nevertheless remains important, he insists, that academic scientists be "objective." "We have right-wing fanatics and left-wing fanatics on all questions," he says, "but scientists should not get involved in those extremes. I get support from conservation groups and from industry, but I make no promises at the outset." Objectivity for Klaine means that the data he produces will not be tailored to serve a decision down the line. Although he refuses to draw a definite line between applied and basic research—"couched within applied research, I ask basic questions dealing with fundamental processes, like how a chemical is degraded by a microorganism"—the construction of objectivity offered by his colleagues who deal more directly with theoretical questions is clearly different from his own. The problems the basic researchers hope to solve are developed and certified within the confines of theoretical science, while the problems Klaine's data will eventually solve arise from life in the muddy world of agriculture, industry, and politics.

Thus for basic researchers, objectivity means distance from general human interest; for applied researchers, objectivity means the refusal to privilege one human perspective over another in advance of the research act. This distinction has a profound effect on the kinds of literature produced by the different approaches to research. Whereas basic research produces strongly argumentative discourse and conclusions designed to clinch a theoretical point, applied research produces open arguments, leaving the conclusions to engineers and policymakers "down the line." As Mark Hinman, Klaine's former graduate student and coauthor points out, moreover, applied research also supplies data for the makers of theoretical models, particularly computer models of ecological processes. Several modelers have asked to use the data from Klaine's project to calibrate and verify their models.

There exists, therefore, a cycle of data usage, with theoretical science feeding methods and interpretations to applied science, which in turn provides data that may be plowed back into theoretical models.

Hinman thinks the distinction between basic and applied research is overstated and is largely a matter of disciplinary politics. He says an "elitism" prevails in efforts to make the distinction stronger than it really is. The aristocracy of basic researchers in natural science prohibits the people best qualified to "make a difference" in environmental matters from doing so: "A big problem with biologists is that they won't go out on a limb and make predictions. I think it's time we trained a generation of scientists to do the job done now by engineers and others who don't understand biology." Certainly science educators who write textbooks and the renegade scientists who use the mass media rather than the refereed journals to publicize their findings would agree with Hinman; they appear to have flaunted the scientific conventions for constructing objectivity.

Hinman himself has not yet gone this far. For him, to be objective means "sticking close to your data." "You have to believe your data to arrive at a better picture of what happens in the real world as opposed to what people think will happen. It all comes down to real numbers, the hard data." The devotion to instrumentally produced data—a "real world" constituted by a mathematical model calibrated and verified by inputs from observations in the natural world—aligns basic and applied researchers and creates their political position in society. Their politics is a defiant refusal to submit to the foregone conclusions of industry or government. One of the reasons that basic researchers look with suspicion instead of (or in addition to) envy upon the well-funded work of an applied researcher like Stephen Klaine—clearly evident in the labs packed with equipment surrounding his office, the swarms of graduate students, the invitations to speak around the world, the smiles of university administrators counting the overhead dollars generated by his research—may well be that the basic researchers, competing for tighter dollars, may view these advantages as the benefits of submission to the needs of extrascientific society. The politics of independence nevertheless persist as strongly among the applied researchers we interviewed as it does among the theoreticians.

From our perspective, the main differences between basic and applied research lie in the rhetorical approaches to their materials, the tendency to develop closed, tightfisted arguments in basic research as opposed to the tendency to produce open-ended arguments in applied research. This rhetorical difference amounts to a distinction of genre. Another such distinction emerges in scientists' use of the jargon term "gray literature."

Gray Literature vs. Refereed Literature

Gary Wein may well be one of the new breed of problem-solving biologists that Mark Hinman envisages.[2] Though he is very much aware of the distinctions that separate basic science, applied science, and engineering, Wein has found himself crossing these boundaries at nearly every stage of his career. Trained as a basic researcher in biosystematics and later plant ecology, he went to work before he had finished graduate school as a consultant for a company called Princeton Aqua Science in southern New Jersey. He did surveys and wrote environmental impact statements involving wetland delineation, wetland evaluation, and wetland creation design. He had in effect become an environmental engineer. From this job, he went on to become the Research Manager for the Savannah River Project in South Carolina, serving as a liaison for research scientists in an ecology project associated with a nuclear reactor facility operated by the Department of Energy, helping researchers give bureaucrats "what they wanted" and at the same time helping the scientists to do "real research." He also did more wetland design and some research of his own.

Several of the papers he produced during his tenure at the Savannah River Project fall into the category of "gray literature," a term frequently employed in the shoptalk of scientists. Essentially it refers to papers written by scientists but not published in refereed journals and therefore not certified by the standardizing protocol of mainstream science, either basic or applied. Gray literature includes government documents, open-file reports, and in-house documents of various kinds. These documents may range widely in their conclusions from basic to applied science and on to engineering, though they are usually devoted to solving definite problems in human life and are paid for by consulting fees or straight salaries rather than by funds from university budgets or research grants.

The following is an abstract from one of Gary Wein's reports that he classifies as gray literature. It deals with wetland design and reconstruction. The paper has clear affinities with applied science, but it differs significantly from the applied studies we have seen that were published in refereed journals. Most scientific readers would see the work as an engineering or technical report:

> The history of a large scale mitigation project of a cooling reservoir (L-Lake) for a reactor on the Savannah River Plant, South Carolina, is presented. The National Pollution Discharge Elimination System permit for thermal effluents discharged into the reservoir requires establishment of a balanced biological community (BBC). As a good faith effort toward establishment of a BBC, wetland/littoral vegetation was planted along 427 m (14,000 ft.) of shoreline in 1987. Approximately 100,000 plants were transplanted. Species planted were representative of submergent, floating-leaved, emergent and woody zones found in regional South Carolina lakes and reservoirs. The transplants have been growing well and reproducing, but project success will not be determined until spring and summer, 1988. (Wein, Kroeger, and Pierce 206)

Like papers in applied science, this one presents an argument that is open-ended, but for different reasons. The applied scientist leaves an argument open as a precaution of objectivity, as a protection against allowing the interests of an extrascientific perspective to intrude upon the data. Wein and his colleagues leave their argument open because the project is not yet finished; their paper is a "progress report" generated by the demands of the bureaucracy that is funding the work. The intended audience is clearly not a group of scientists, but rather funding managers and other government officials, whose lack of expertise is indicated by the need to have metric measurements translated for them. The very impetus to write is extrascientific; the purpose is to keep the money flowing. Moreover, the report is laced with political interests, claims of behaving in "good faith" and according to accepted laws, policies, and procedures.

And Wein is no stranger to environmental politics. Despite his background in science, his years of working in government and industry have led him away from the politics of apolitical research. In another paper about his work at the Savannah River Project, published in a book on wetland resources but still not subject to the kind of full review expected in mainstream science, he uses judgmental adjectives and a rhetoric of recommendation that is utterly foreign to academic science. Here is the abstract:

> Three factors that can increase the complexity of the wetland mitigation process are 1) scale, 2) multiple regulatory agencies, and 3) conflicting goals and objectives. Scale includes the physical size of the project as well as the magnitude of alterations to ecosystem processes. The mitigation process involved in the restart of a nuclear production reactor is used to illustrate how these three factors could increase mitigation complexity. Two mitigation components of the reactor restart are presented. One component, the establishment of a balanced biological community in L-Lake, had large-scale impacts, nebulous goals and objectives, and involved several regulatory agencies. In contrast, the second component, replacement of lost foraging habitat for the endangered wood stork at Kathwood Lake, had small-scale impacts that could be managed, a singular, well-defined goal, and clear guidance from one regulatory agency. The L-Lake project has yet to succeed, while Kathwood Lake has been immediately successful. (Wein and McCort 41)

The rhetoric of this paper shares much with the rhetoric of basic research. In its review of two experiments, it clinches a tight argument about which of two kinds of project—the one large-scale and diffuse, the other small-scale and closely managed—is most likely to succeed. But the conclusion is oriented toward policy rather than toward theory and is therefore worlds away from basic scientific research. We asked Wein about the use and interpretation of scientifically generated information: When can we say that such data are no longer used "scientifically"? "When you use that information as a lever to get the DOE to change their actions," he responded, you have moved out of the realm of mainstream science.

This response affirms what we found again and again in our interviews with academic scientists. If as in Walter Beale's model of the discourse aims, the construction of scientific discourse as primarily contemplative is constantly undermined by the intrusion of the rhetorical, instrumental, and poetic aims, we must nevertheless realize that scientists themselves distinguish their own rhetoric and their own instrumentalism from that of the world outside the scientific community. When science begins to influence the rhetoric and the instrumental realities of the public realm, it loses its special character and becomes something other than science as defined by scientific authorities. The author who makes recommendations for actions based upon scientific data or who insists upon the factual status of information still controversial in research circles is no longer doing science but is rather using science in a distinctly public rhetorical appeal.

Contrasts: The Discourses of Research, Pedagogy, and Public Action

We have now shown that scientifically certified discourse is different from discourses like scientific activism, which appeal to science without doing science, and that it is structured to resist the intrusions of general politics and human interests. The French philosopher Jean-François Lyotard, in his book *The Postmodern Condition*, usefully conceptualizes the difference between the esoteric discourse of science and the public version of that discourse. Since this analysis is consistent with our view of discourse communities, we might dwell for a moment upon it.

Narrative and Scientific Discourses

Lyotard begins with a distinction between what he calls narrative discourse and scientific or "denotative" discourse. Narrative discourse deals with ordinary human actions in a communal setting. Narrative is the standard means of gaining knowledge in folk communities. Like all discourse, it involves a speaker, a listener, and a subject matter, the latter being always a hero or an heroic action. The speaker's only claim upon the right to speak is that he or she is a member of the community and has something of interest to say. The speaker may have been the hero of the action ("I did so and so"), but not necessarily.

The speaker needs only to have heard the story ("I heard yesterday that . . ."). So after the listeners hear the story, they are certified to become speakers in turn. The participants in the discourse act—the speaker, the listeners, and usually the hero—are known to one another, are familiar with the conventions of narrative, and are empowered to participate in the discourse merely by their membership in the community.

Scientific discourse, on the other hand, is denotative rather than narrative, according to Lyotard. The functions of speaker, listener, and subject matter are pragmatically distinct. The subject matter is, at least theoretically, determinable equally by all participants, but the speaker (or writer) asserts a special claim on the truth and is able to furnish proofs of that truth. The listener is, in theory, the equal of the speaker but can become so in practice only by being able to provide the same kind of proof provided by the speaker. How are these proofs determined to be good proofs? They are established by the rules of the scientific community, by methods and theories approved in advance and agreed upon by a consensus of experts.

Scientific discourse is like narrative discourse in that it requires a communal context, a culture whose rules govern contributions to the discourse. The interchangeability of the speaker and listener are also similar, but here differences begin to emerge. Narrative creates its truth; denotative discourse establishes truth through method. Method requires mastery. A master speaker requires a master listener. To become either requires a teacher.

Thus science divides into a research function and a teaching function. Research is denotative and disputational, a discourse of proofs and challenges, formally arranged and carefully structured. Teaching, on the other hand, provides the occasion for the reemergence of narrative within science. Master researchers tell stories that gradually bring student researchers up to the level of full participation in the discourse community. To perform their narrative function, members of the community take some liberties with the truth. To learn the method, students must start with some statements that are "transmitted through teaching . . . in the guise of indisputable truths"—*the facts* (Lyotard 25). Students, the untrained listeners (as well as teachers not themselves a part of the research community—secondary school teachers, for example), tend to grasp at the facts of science as a secure ground in the whirl of method and disputation. Only as they approach the level of the master researcher are these students able to treat the facts as black boxes that the effective application of method may eventually open again, producing controversy and perhaps even falsification. In general, research scientists and theoreticians consider simple narrative discourse with its solid facts to be contemptible, primitive, unsystematic, and underdeveloped (Lyotard 27). They must, however, tolerate a measure of this kind of discourse in order to reproduce their community of speaker-listeners of more or less equal status.

From the perspective of the research community, the trouble with writers like Aldo Leopold, Barry Commoner, and Rachel Carson is that they perform in public the function of scientific pedagogy, thus extending the function beyond the simple needs of the community to reproduce itself. Their narrative discourse—Leopold's story of the philosophical blacksmith, Carson's fable of the blighted town, Commoner's explanation of the facts of entropy and the causes of Lake Erie's near death—strike out into the political arena with stories for listeners without the special expertise needed to question the assertions of fact. A little knowledge is a dangerous thing. According to the conventions of narrative, these listeners may themselves become speakers in the extrascientific community, armed now with knowledge that bears the esoteric inscription of science: "I heard from a reliable scientist that. . . ."

Metanarrative

Once released into the public at large, the pedagogical discourse of science comes under the influence of another form of discourse, which Lyotard calls *metanarrative.* Metanarrative is distinguished both from the plain narratives of folk communities and from the denotative discourse of science. Its purpose is hegemonic; it aims to bring other discourses under the control of its broad explanatory power and thereby to influence the use of such discourses in a community that includes smaller communities—such as the Roman Empire, the university as it has developed over the last one hundred years, or the global mass public created by modern communication and transportation technology.

Metanarrative assumes the general shape of the simple narrative but shares with science the concern with legitimation. The metanarrative of Enlightenment, for example, evolved in nineteenth-century Germany into an effort to bring the diverse elements of learning into the project of an all-encompassing knowledge, symbolized by the invention of the university. All areas of learning, the story goes, seek ultimately the same Truth, whose metanarrative is produced and certified by philosophy—hence the grouping of all specialized research under the degree called Doctor of Philosophy.

Likewise, the metanarrative of human liberation has dominated political ideology at least since Hegel's time. This is the story of progress toward an ever broader human liberty—the freeing of slaves, followed by the emancipation of women, ethnic minorities, workers, and so on. But the range of the metanarrative is rarely limited to the political or social fields. Psychology, for example, has adapted this metanarrative to the story of a person's "state of mind"; in waking life, the id is said to be encumbered by the ego while the ego itself is enslaved by the superego; but in the dream "state" or in neurotic "states," the repressed forms of mental activity are liberated. The discourse that promises the greatest progress in the liberation of the world spirit is that which is the most legitimate in the metanarrative of human liberation.[3]

The metanarrative that prevails in American political (and psychological) life is the ideology of progress (which includes apsects of both Enlightenment and human liberation). Ecospeak may be seen, for example, as an effort to take environmentalism—which is actually a broad social concern, a loose collection of little narratives—and reduce it to the status of a protest movement, a rear guard attack on developmentalist progress, a regression into primitivism. Environmentalism may also be interpreted (in the manner of Hays) as an avant-guard movement demanding the extension of social progress, represented by "environmental amenities," to ever greater numbers of people. Along these same lines, environmentalist progress may be aligned directly with the metanarrative of human liberation. For example, the eco-anarchist Murray Bookchin, following Lewis Mumford, has argued that "the very notion of domination of nature by man stems from the very real domination of human by human" (*Ecology of Freedom* 1) and that "*all ecological problems are social problems*" (*Remaking Society* 24). In formulating the land ethic, Aldo Leopold attempted to extend the metanarrative of human liberation to include the liberation of nonhuman nature from the dominion of humanity. (He nevertheless used Hegelian logic and Hegelian language and therefore worked within the old metanarrative.) The deep ecologists, for their part, have followed Leopold in creating a new Hegelian doctrine of the "rights of nature" (see Manes). They have thereby interpreted anew the idea of progress, positing a planetary progress that may well stand in opposition to the kind of isolated human progress favored by reform environmentalists and developmentalists alike.

To think of political change outside of the metanarrative of progress is very difficult, but to do so, either by creating a new metanarrative (such as Habermas' concept of communicative action) or by abandoning metanarrative (as in Lyotard's postmodernism), may be the step that is needed to overcome the environmental dilemma. Science may be helpful in showing the way beyond. For science, as for many folk communities that perpetuate themselves through diverse narratives, progress is largely irrelevant. The main thing is to *sustain* the scientific research project or the folk community. If we could stop thinking of science as merely a data base for bolstering preformed arguments about environmentalist or developmentalist projects, and if we could consider the *form* of scientific discourse as a model of success, then new paths might open beyond the tangle of our intractable problems. That form consists of an informational system that is open to the extent that it allows for falsification, even whole paradigm shifts, but closed to the extent that it resists being appropriated by foreign metanarratives.

An environmentalist discourse has begun to develop along these lines in ecological economics, in which the paradigm of *sustainability* has evolved as a challenge to the story of economic growth that proceeds in an never-ending upward spiral. Sustainability resists this economic version of the metanarrative of progress and thereby promotes maintenance and endurance of the ecological systems on which human life depends. This project, which we will consider in chapter 7, involves a serious rethinking of the idea of progress and potentially breaks the hold of ecospeak.

But first, we must consider other efforts to bring scientifically generated information to bear on public understanding and environmental policy. In chapter 4, we turn from the denotative discourse of mainstream scientific research and from the narrative of science pedagogy, to science news, an attempt to create for a mass culture the equivalent of the culture-sustaining narratives in folk communities. Though it deals with the subject matter of science, this narrative differs from both refereed literature and gray literature, both of which require a writer who is ostensibly a member of the research community; it also differs from narratives constructed for the purpose of scientific pedagogy, which requires the pragmatics of a master addressing novices in a specialized knowledge. In science journalism, a writer, reporting to an audience of equals, tells the story of heroic actions with great import for the future of the community. This mythologizing of scientific learning empowers the listeners to become speakers themselves: "I read in *Time* magazine that the earth is warming up because of pollution. . . ."

It is ordinary people, people "just like us," who take part in public and social life, endowed with intellectual faculties, feelings, drives; therefore it will be appropriate to intervene in this endowment with a global and comprehensive set of tools. This is why rhetorical discourse must simultaneously pursue and bring together three different goals: *docere*, teaching on an intellectual level; *movere*, touching the feelings, the emotional "experience" of the audience; finally, *delectere*, keeping their interest alive, soliciting their attention so that they will follow the threads of one's thoughts, without becoming bored, indifferent, distracted.

—Renato Barilli,
Rhetoric (ix)

The public's perceptions, of course, are primarily shaped by . . . the media, which thrive on the four D's: drama, disaster, debate, and dichotomy.

—Stephen Schneider,
Global Warming (206)

4

Transformations of Scientific Discourse in the News Media

Scientific Facts, Human Interest, and the News

News reporters and the purveyors of mass media have an ostensible commitment to a realist, even a positivist, epistemology. They are supposedly devoted to the facts. In principle, their brand of objectivity resembles that of applied science, a hardheaded insistence on maintaining their own perspective against the pressures and intrusions of governmental and corporate powers.

Even the quickest analytical pass at the discourses of applied science and mass journalism, however, will reveal divisions that make this kind of comparison seem trivial. Above all, scientists who write and read denotative discourse based on personal involvement with experimental data are less dependent on uncertified secondhand information than are journalists, most of whom depend exclusively on interviews (micronarratives); and if a particular source proves to be more willing about sharing information or to have a more interesting slant on a particular story, journalists may consciously or unconsciously privilege that source and thereby betray their own objectivity. Moreover, the frantic pace at which news reporters work, unlike the more leisured pace of scientific data collection, makes it difficult for the journalists to get sufficient distance on a story before they go to print. Then there is the much-discussed problem of time and space limitations in the mass media, which fail to allow for sufficiently developed stories on even relatively uncomplicated issues, much less the complexities of environmental science and politics.[1]

But even more important than these differences in the way information is generated in the two fields are differences stemming from the journalists' understanding of information value, which ultimately ensure that the facts of science will be distorted or reinvented altogether when they are presented in the news media.

Information Value: The Concepts of "News" and "Human Interest"

The mass media's interpretation of information value rests upon two related conventions of journalism—the concept of "news" and the concept of "human interest." For a story to be considered "news," it must tell readers something they don't already know, something they haven't already heard or become accustomed to. News dwells upon the unfamiliar, the strange, the huge, the surprising turn of events, the trouble spot, the crisis. In this sense, news reporting is the rhetorical equivalent of crisis-based government. A story doesn't "break" until its effects are already dramatically evident and are certain to have a broad impact. And once it has broken, it is no longer news; thorough follow-up reports are rare unless the effects of the crisis continue to make themselves felt in a dramatic manner. A shipwreck that causes millions of gallons of oil to be spilled in Alaska is news even if that spill represents only a small fraction of the total amount of oil leaked into the oceans every year. Smaller spills are weekly occurrences. They are nothing new, so they are not news—except perhaps as background to the larger crisis that commands media attention. Similarly, the daily ploddings of the scientific laboratory, the adjustments of theory, the small but significant contributions to understanding cannot be reported as news unless they lie along the trail to a "breakthrough," a big science story. And what makes a big science story?

Human interest is the leading factor in determining what scientific activities will be covered as big stories. This approach may occasionally involve journalistic portraiture, especially on the local level; a scientist taking an unusual approach to an old problem or one who has won a large grant for research may become the subject of a back-page feature story. By and large, however, there is a strong tendency first to focus on applied research—especially research connected with human problems that are already established as "issues" worth dealing with—and second, to confuse science with engineering, a discipline structured entirely around solving human problems. Some magazines, the ones devoted exclusively to science journalism—*New Scientist*, for example—or prestigious journals like *Science* and *Nature*, which have sections for science news, report on a broader range of scientific activities and do cover theoretical developments. Yet even in these journals, choices must be made about what to include, for there is a lot of science in the making and not all can be covered. Thus the tendency to focus on applied science and issue-oriented questions prevails.

The human interest approach goes against the grain of scientific objectivism for two reasons. First, it insists that science must have social value outside of its own pursuits, that science cannot dwell within its own history but must make connections with the larger history of humankind, that science cannot be an end in itself; all science must be applied science. In this sense, the mass media join government and corporate organizations in the institutional intrusion upon the autonomy of scientific investigation. Second, this approach insists that science must not only be applied to general human problems but must also press toward the kinds of conclusions that are generally absent in typical papers reporting findings in applied research; science must *solve* human problems and thus must transcend its own version of objectivism, its own self-definition, must become engineering if it is worthy of being reported in the press.

We mentioned in chapter 3 that no working scientist would accept a discourse taxonomy that grouped science journalism with scientific writing. It is precisely this emphasis on human interest and the insistence that what is reported be "news" that puts distance between the journalistic and the scientific outlook. The emphasis on human interest carries the journalist out of the field of natural science and into the action-oriented fields of social movements and politics. Moreover, the striving for the dramatic and even the sensational slant on the facts implicit in the demand for "news" causes the genre to shade into the field of poetic or mythic utterance. News reports do provide information, but they must do it in a form that is engaging, entertaining even, if they are to sell papers and magazines, which in turn sell the advertisements that pay the bills. The result is the making of science into a story with implications far beyond those of the laboratory and the library, a story tied to the longing and struggles of humankind in general, a veritable mythologizing of science.

When the issues of "warm science," scientific research and theory that have not yet attained the cold solidity of fact, are brought into contact with the overheated rhetoric of public debate on issues like environmental degradation, the conditions for controversy are complete. By 1988, the importance of environmental issues was certified by a New York *Times* poll, which showed that "80 percent of the public agreed with the statement, 'Protecting the environment is so important that requirements and standards cannot be too high, and continuing environmental improvements must be made regardless of cost' " (Suro). When the press got wind that scientific theorists were pursuing an explanation of why the summer seemed hotter than usual, an explanation that connected climatic change to technological damage to the environment, a media blitz was almost certain.

The Case of Global Warming

The coverage of research in global climate change during 1988 and early 1989 has all the elements of high journalistic and scientific drama. Above all, it demonstrates that the scientific perspective of objectivism must be radically denatured if it is to pass muster as a force in journalism or in public policy making.

The hypothesis of the "greenhouse effect" has been around for several decades now, long enough to have received full treatment in recent textbooks (see for example, ReVelle and ReVelle). It has become a well-established theory in atmospheric science and is widely used, for example, to explain why the surface temperature of Venus is so much higher than that of Mars. Venus has an atmosphere rich in carbon dioxide and other gases and particles that act like the glass panes of a greenhouse; they trap the sun's radiant infrared energy so that more heat remains inside than is allowed to escape. The atmosphere of Mars, on the other hand, is poor in carbon dioxide and thus releases more of the sun's heat back up into space. While the Martian surface is frigid, the atmosphere of Venus is a "runaway greenhouse," far too hot to support life (Schneider, "Greenhouse Effect" 771).

The controversy arises when the theory is applied to human influences upon the earth, when it is suggested that our activities have so unbalanced the earth's carbon cycle that the character of the atmosphere is undergoing a change that will have catastrophic effects on the global climate. As early as 1827, the French scientist and mathematician Jean Baptiste Fourier suggested that human activity might alter climate; and in 1957, in a profound overlaying of the scientific perspective on human history, Revelle and Seuss coined what has become a very popular phrase among scientists, "large-scale geophysical experiment," to describe the tendency of human-influenced processes to result in an increase of carbon dioxide in the air (Ramanathan 293–94). Normally, a balance is maintained between the introduction of carbon dioxide into the air and the incorporation of carbon dioxide back into the global ecosystem. In the oceans, carbon dioxide is dissolved; in green plants, it is used in photosynthesis, which releases oxygen. But processes like electric power generation, industrial combustion, and the burning of fuel for transportation, cultivation, and the heating of homes and factories have overloaded the atmosphere, releasing more carbon dioxide than the natural processes of conversion can accommodate. The increased level of carbon dioxide in the air produces the greenhouse effect. Eventually the temperature may stabilize, but at some higher level. According to the grimmest forecasts based on the greenhouse hypothesis, these atmospheric alterations could result in a net gain in global warmth large enough to cause substantial melting of the earth's polar caps and a consequent rise in sea level, making important population centers (New Orleans and the Netherlands, for example) uninhabitable, thereby exacerbating the problem of overpopulation and famine, which social scientists already rate as a crisis. The worst-case scenario suggests that in areas with the most fertile soils—the American Midwest, for example—decreased rainfall and increased heat could lead to desertification, while temperate climate patterns could prevail in areas with soils less inviting for large-scale agricultural development (New Mexico, for example). There could also be negative contributions from related problems, such as acid rain—also caused by emissions of waste gases from industrial and technological processes. Faced with these possibilities, a number of scientists have become concerned about how soil and water fit for human and animal use can be sustained.

These scientists—most of them specialists in atmospheric science, like Stephen Schneider, James Hansen, and George Woodwell—claim that there is a broad consensus among scientists that the earth is warming up. But few will claim consensus about the causes, the rate, or the future of the warming. The dominant outlook in scientific discourse about the application of greenhouse theory to problems involving the human relation to nature therefore remains cautious overall. A typical argument is that not enough is known to make solid conclusions. This general viewpoint is adequately expressed by the University of Chicago atmospheric specialist V. Ramanathan in an article in the 15 April 1988 issue of *Science*: "Observed records do reveal a warming of the order of about 0.5 K, but the temporal history of the warming is unlike the pattern anticipated by the theory. The warming occurred abruptly and in bursts. Either the observed warming is not related to the increase in the trace gases or current theories and models of ocean-atmosphere interactions are inadequate to capture the transient response to the climate system to a time-dependent variation in external forcing. The theoretical understanding of the climate system is by no means complete" (298).

This very article, however, was likely published as part of an increasing interest about the greenhouse theory among scientists as well the general public. And several well-respected scientists believe that global warming is due to the greenhouse effect stimulated by human alterations of the atmosphere. James Hansen of the Goddard Institute for Space Studies, in a testimony before the Senate Energy Committee on 23 June 1988 reported that the year's hot temperatures and drought were "99%" likely to be a result of greenhouse warming (qtd. in Schneider "Greenhouse Effect" 779). And the respected atmospheric specialist Stephen Schneider felt strongly enough about global warming as a general human problem that in 1989 he issued an "instant book," *Global Warming: Are We Entering the Greenhouse Century?*, written (in four months) on the model of *Silent Spring* and published by Sierra Club books.[2]

In the 1980s, three factors have increased concern about the possibility of human-induced climate changes and have led more scientists and other observers to take the greenhouse hypothesis seriously:

- There is mounting evidence that chloroflourocarbons (CFCs) and other complex organic gases released during industrial processing intensify the effect of carbon dioxide and other "greenhouse gases."
- There is also evidence that these gases may contribute to the destruction of the ozone layer in the stratosphere, one of the important chemical buffers that moderates the intensity of the sun's rays in the lower atmosphere. In 1985, a previously unrecorded "ozone hole" was discovered over Antarctica.
- In recent years, a pattern of galloping desertification has developed in many areas of the world, which when coupled with the destruction of rain forests for cattle production and other forms of development, clearly weakens the ability of the earth's green systems to assimilate carbon gases.

With these developments in science and social science in the background, the hot summer of 1988 brought to the United States the worst drought in recent history. The Midwest, the "world's bread basket," was hit worst. The mass media, which had formerly shown only passing interest in the greenhouse effect, took up the topic with fervor. When tied to the human interest in weather and crop production, global warming attained the status of a news sensation.

Rhetorical Range in Science News Coverage

Unlike some analysts of public information, we do not mean to suggest that the news media have developed into a monolithic institution governed by absolute laws (an analytical oversimplification that mars Schneider's otherwise valuable critique in *Global Warming*, for example). Indeed, in both newspapers and magazines, there has been a significant rhetorical range in the coverage of global warming. Despite the widespread commitment to the concepts of news and human interest, differences in target audiences generally have strong effects on the discourse of science journalism.

In July of 1989, we analyzed various reports on how scientific findings relate to the potential crisis in climate, focusing on the the timing and number of the reports and the tone of the reports, with special attention devoted to the image of science emerging from them. In the coverage of global warming in the weekly press, we found that, in matters of timing and number of reports, the magazines followed very similar patterns despite other differences. The number of stories on the greenhouse effect increased dramatically around the time of the hot summer of 1988. The general tone and the image of science cultivated in these reports, on the other hand, varied widely across types of magazines, as distinguished by their target audiences.

To give a sense of this range of coverage, we will present an analysis of articles appearing in two widely different magazines—*Science*, whose target audience is mainly composed of working scientific researchers, and *Time*, whose target audience is the general public.

Science is a prestigious scientific journal that in addition to publishing refereed papers in basic and applied science, with a preference for research in the life sciences, also publishes science news reports and essays. The highly technical "reports" are written by and are meant to be read by specialists in such fields as biology, chemistry, and atmospheric science. But the stories in the sections called "research news" and "news and comment" are written by staff reporters with scientific training in the area covered. Staff writer Richard Kerr, for example, has specialized in atmospheric science and has written the great majority of the news articles on global warming and related issues in *Science*. Despite this concession to specialization, however, the articles are intended to be accessible to readers who, though acquainted with general scientific principles and reasoning, are perhaps not familiar with the specialized methods and recent findings on the topic covered. A third group of writings is called "articles"; these are written by noted specialists in the field, and though intended for a general readership, are more challenging than the "research news" and "news and comment" stories. Because it addresses primarily an audience of scientists and prints articles written either by active researchers or by journalists with scientific training, *Science* conveys an image of scientific practice and a tone that scientists themselves would likely be most comfortable with. We hypothesized that it would also be less concerned with topics dictated by extrascientific historical forces, but we found that although it has reported on the development of the greenhouse hypothesis since its inception, the number of reports increased substantially in the late 1980s, and the topic became a major theme for both its news reports and its technical articles in 1988. Thus the timing and the number of the reports were certainly influenced by many of the same factors that determined the content of magazines written for a nonscientific audience. Human interest is clearly a major force in the selection of what constitutes reportable science news. Still, the overall tone of the journal tends toward objectivism and at times, irony. It is, in short, *insider journalism.*

Time magazine enjoys one of the largest circulations of any weekly magazine in the world. It regularly reports on human interest topics in science and in recent years, has even included a section on "the environment." Because it treats such a variety of kinds of news, however—national and world politics, domestic concerns, economy, entertainment, and so on—its coverage of news about environmental and other scientific topics has been, until recently, highly selective. *Time* is prototypically journalistic in its concern with human interest and the news. It portrays what scientists often disparagingly call the "popular image of science," preferring applied research and engineering to theoretical concerns, and wavering between reverence and mistrust in its portrayal of the esoteric knowledge of scientists. Its preferred tone is one of urgency, a tone that rightly accompanies reports on crises. Its stories on global warming and other environmental issues in late 1988 and early 1989 seem to have come out of nowhere and were no doubt stimulated almost exclusively by the events of the summer of 1988. The traditional "Person of the Year" cover story for the first issue of 1989 was devoted to the "Endangered Earth" in an unprecedented "Planet of the Year" issue. Environmental reports have been offered almost every week thereafter. Like *Time, Newsweek* and other popular weeklies have devoted cover stories to the environment. The topic of environmental degradation has become a certified *issue,* largely because it is perceived as having crisis potential.

Irony and Community Formation in *Science*

No doubt the scientific community has a strong ecological conscience—a conscience reflected in the editorials of *Science,* one of the leading magazines in that community. In an editorial of 3 July 1987 entitled "Inexorable Laws and the Ecosystem," Donald E. Koshland pulls out the rhetorical stops in a plea for global approaches to stronger controls on population, land use, and pollution, urging developed nations (like the United States) to set examples through self-regulation and not to expect developing nations (like Korea) to have the same economic capacity to control their own growth. Much in the manner of Barry Commoner, Koshland paraphrases the third law of thermodynamics in a colloquial sentence, "You can't get out of the game," and applies that law to the global environmental problem: "We have only one atmosphere whose balance between carbon dioxide, oxygen, and ozone is critical to our survival. We have only one earth and ocean whose fertility and purity are equally important" (Koshland 9). In the editorial section of *Science*, the scientist-reader is appealed to as a human subject who has temporarily stepped out of his role as objective researcher.

The journal's rhetoric undergoes a strong shift, however, as soon as we move from the editorials into the "Research News" section. Here the reader is not expected to be a specialist, but is expected to read as a fellow researcher with a deep respect for the values of the scientific community, above all the commitment to the paradoxical politics of noninvolvement.

Defying the Conventions of News and Human Interest

In a 9 October 1987 report on the Antarctic ozone hole, Richard A. Kerr provides a typical treatment, maintaining a reporter's objectivity while giving a full sense of the rhetorical interplay among the various hypotheses active in this "warm" issue:

> Scientists returning from instrument-laden flights into the thinning ozone layer over Antarctica, the worse thinning seen so far, say they have support for both sides of the debate over the hole's cause. For atmospheric chemists who argue that chloroflourocarbons form the hole, the chemistry encountered in the Antarctic lower stratosphere was wildly perturbed, much as predicted. Man-made pollutants would appear to be destroying ozone every austral spring.
>
> For meteorologists who suggested that the hole could merely reflect the unique weather of the Antarctic stratosphere, the hole-pene-

> trating flights showed that the extreme cold there plays an essential role in driving ozone destruction by forming ice particles that help accelerate the chemical reactions. Because such conditions are rare or intermittent at lower latitudes, it could be argued that the chemical destruction of ozone outside the springtime Antarctic is not likely to be proceeding faster than conventional theory predicts.
>
> However, chemistry cannot explain completely the sudden deepening of the hole since about 1976 or springtime decreases that extend halfway to the equator. These ozone losses probably involve changes in the winds that carry ozone toward the pole. Thus, purely meteorological influences are also thinning the ozone layer. ("Winds, Pollutants" 156)

Kerr willingly enhances the drama of warm science with an artful use of modification and colorful diction in general ("the worse thinning seen so far," "wildly perturbed," "hole-penetrating flights"). In a report published in the middle of the 1988 summer, Kerr's language is even more conventionally dramatic. He begins this way: "As Earth stands on the brink of a global temperature increase unprecedented in the history of human civilization, the international scientific and policy communities are mobilizing to minimize the effects of global warming. Scientist's views of the future are as murky as ever, but there is a new sense of urgency, fueled in part by disquieting surprises in the stratosphere" ("Report Urges Greenhouse Action" 23). And consider this passage from an article written a month later by the same author: "The Antarctic ozone hole was a startling and ominous discovery. Theorists had not seen it in their maze of equations predicting the chemical behavior of stratospheric gases. And high-tech sensors in space picked up the hole, only to have that data expunged by computers as too incredible. Once recognized, the hole and the ease with which it consumed ozone were disquieting, to say the least" ("Ozone Hole Bodes Ill for the Globe" 785). The frequency in these passages of adjectival and adverbial modifiers, the traditional vehicles of judgment and subjective assertion in English syntax, is far greater than we might expect to find in writing that is crafted to appeal to objectivist researchers. But Kerr uses such language mainly to reflect the warmth of the debate rather than strongly to favor one side or the other. A close rhetorical reading of the opening passage from "Winds, Pollutants Drive Ozone Hole" quoted above might suggest a slight privileging of conventional meteorological theory, evident in the tendency to place the presentation of conventional arguments after the interpretation of ozone thinning as caused by pollution, thus allowing the well-established hypothesis to have the last word. The same organization is preserved in the report as a whole. This tactic is consistent with the overall "show me" attitude of scientific conservatism, a Popperian outlook that demands that new theories work hard to displant, to "falsify," old theories.

Moreover, using a tactic that represents a step away from the conventions of human interest reporting, Kerr places his emphasis on the results and conclusions of the research rather than on the researchers themselves. In ignoring names and biographical details in favor of data and conclusions, he defies the journalistic trend to focus on personalities, a trend best observed, for example, in the work of John McPhee or in Jonathan Weiner's book *The Next One Hundred Years*, which develops the history of research on global warming in what amounts to a group biography of the major researchers. In contrast, Kerr offers no clear images of the key actors in the drama of his story. In "Report Urges Greenhouse Action Now," we find an extreme example of this approach in the singling out of a renegade scientist, probably James Hansen, though we cannot know for sure since *the name of the scientist is never mentioned*: "The U. S. droughts and the century-long global warming culminating in the 1980s . . . are catching the public's attention, despite scientists' refusal to link any one climate extreme to the greenhouse. Even the claim by *a lone expert* that the greenhouse has arrived has failed to gain support from other scientists" (23; italics added). Without the usual passive voice, indirect sentence beginnings, and other conventions of impersonal scientific writing, the prose still avoids an overt humanization of the scientific process. No heroes or villains are created, no praise or blame bestowed. Scientists are portrayed as a closely knit communal group, and a conservative one at that.

Skepticism and Open-Ended Narratives

Thus despite the use of dramatic language, Kerr's articles in the "research news" section of *Science* preserve the trend toward skeptical caution, conservatism, and impersonality implicit in the scientific construction of objectivity. Measured by the popular standards of entertainment value, this rhetorical approach is decidedly anticlimactic. The reading public is usually thought of as desiring a clear answer, a definite end to the story. But science is not designed to provide such melodramatic closures. As a consequence, public interest in science is very difficult to sustain. As one commentator has suggested, "The public has become used to conflicting opinion on health issues and position reversals on topics ranging from dietary recommendations to the depletion of the ozone layer. Many have come to feel that for every Ph.D. there is an equal and opposite Ph.D." (qtd. in Goldsmith 17).

Reporters in *Science* routinely ignore the general public's taste for dramatic constructions with a clearly defined beginning, middle, and end—stories in which someone wins and someone loses—in favor of the more open-ended stories in which the conflict ends with no clear victor, or in which the conflict produces one small adjustment that opens the way for others, or in which a well-trod path is closed without a new one having yet been provided. In an article published in the February before the eventful summer of 1988, Kerr poses the question that would receive such wide coverage a few months later: "Is the Greenhouse Here?" Once again, in a review of the recent findings of greenhouse investigators around the world, he unpacks the communal image of science as cautious and skeptical in the very first sentence: "Proving that increasing concentrations of carbon dioxide and other trace gases in the atmosphere are altering Earth's climate is years if not decades away [*sic*]" (559). In an effort to communicate to the nonspecialist the difficulty of detecting the greenhouse effect, Kerr spins out an analogy that the scientists themselves have used, a comparison of the research with criminal investigation:

> No one type of change is likely to convince the scientific community of the reality of the greenhouse effect or its true magnitude until well into the next century, when the world could be condemned to dramatic changes. The best bet for early detection seems to be the identification of a number of changes—warmer weather, warmer ocean water, a cooler stratosphere, and increased precipitation, for exam-

> ple—that together would, in all likelihood, be caused by a greenhouse.
>
> The approach now being pursued is more like developing a composite picture of a culprit rather than arresting the first suspect who has the same color eyes. Researchers call the technique fingerprinting, although it is hardly as conclusive as the human technique. And they are beginning to scrutinize what may be the first components, albeit fuzzy ones, of the greenhouse fingerprint. ("Is the Greenhouse Here?" 559)

The pattern of skeptically asserting a bit of evidence and then retreating from it or revealing its shortcomings as evidence—a move paralleled in this passage's offering of an explanatory analogy and then deprecating its accuracy ("Researchers call the technique fingerprinting, although it is hardly as conclusive as the human technique")—is the main strategy for presenting the findings of individual research teams. Here is a typical passage:

> Computer models predict a warming by now of anywhere from 0.3 to 1.1°C, depending on the initial carbon dioxide concentration, the ability of the ocean to slow the warming by absorbing heat, and the sensitivity of Earth's climate system to carbon dioxide, rather than the models'. If one includes the uncertainties inherent in the increasing amounts of other greenhouse gases such as methane and chloroflourocarbons, whose effects could eventually double those of carbon dioxide, the consistency of one type of observation with prediction is not so impressive. ("Is the Greenhouse Here?" 559)

Having thus established a tone of caution and a skeptical rhetoric, the article concludes with a studied anticlimax: "Fingerprinting is only just getting under way, but the chase is on" (561).

Apocalyptic Narratives with Qualifications and Demurrals

Though maintaining the skeptical tone, the articles by Kerr and other writers in *Science* that appeared after the summer of 1988 do pick up a few rhetorical threads typical of the more popular coverage of global warming. In the July 1988 article, "Report Urges Greenhouse Action Now," Kerr reviews a study produced by a group of ten experts from three major international organizations—the United Nations Environment Program, the World Meteorological Organization, and the International Council of Scientific Unions. He carefully distinguishes this effort from normal scientific research, stating that scientific predictions about the future climate, unlike the clear recommendations for action posited by the report, are "as murky as ever." But then he goes on to follow the report in developing a worst-case scenario, an apocalyptic narrative typical of mass media and polemical writings on future ecological trends:

> The . . . report does not draw on any fundamentally new evidence, but it does project the future warming in some novel ways. Instead of emphasizing the oft-quoted 1.5° to 4.5°C global warming expected with a doubling of carbon dioxide, the primary greenhouse gas, the report estimates rates of temperature increase due to CO_2 plus other greenhouse gases. In a worst case scenario, in which greenhouse gas emissions are unrestrained and the climate is highly sensitive, the estimated rate is 0.8°C per decade. . . .
>
> That would . . . carry Earth by late in the next century into a climate as warm as any for hundreds of thousands of years.
>
> The warming could hardly go unnoticed. The sea would warm and expand while glaciers melt, pushing up sea level about 30 centimeters by the middle of the next century and possibly as much as 1.5 meters. Even the more modest rise would erode most sandy beaches along the U.S. Atlantic and Gulf Coasts at least 30 meters. The heat and some regional dryness would disrupt agriculture in some places, especially semiarid regions where agriculture is marginal to begin with. (23)

Parlaying scientific findings into stories of the future, in an effort to shape policy by creating alternative stories, is the primary rhetorical tactic of most policy-oriented reports, such as the one described here. It is a way of drawing out the conclusions of typically open-ended reports on applied research. In contrast to more sensationalistic journalism produced in the tradition of *Silent Spring*, Kerr's rendition of the apocalyptic story is definitely understated. The global warming, he says, "could hardly go unnoticed."

The same is true of other writers in *Science*. Consider this paragraph from a March 1989 report by Elliot Marshall on the Environmental Protection Agency's plan for inhibiting global warming:

> If nothing is done, the resulting temperature increase by year 2100 could be enormous, according to EPA, ranging from a minimum of

> 2°C to 3°C in a slowly developing world to a high of 5°C to 10°C in a rapidly changing world. For reference, the difference between the mean annual temperature in Boston and Washington is only 3.3°C, and the total global warming since the ice age was about 5°C—a change, EPA says, that "shifted the Atlantic Ocean inland about 100 miles, created the Great Lakes, and changed the composition of forests throughout the continent." With a reversal of deforestation, a cutback on fossil fuel use, and a virtual ban on chloroflourocarbons, the impact could be reduced 60%, so that global warming would be no more than 0.6° to 1.4°C. (Marshall 1544)

The whole passage seems built to accommodate a structure of cautious qualifications of the adjective *enormous* in the first sentence. The use of that word is followed immediately by the disclaimer "according to EPA." And the apocalyptic comparison to ice age history is introduced by the rationalizing phrase "for reference," suggesting that the author, well aware of the usual political use of such stories, is claiming that the radically condensed version of the apocalypse—quoted directly from the EPA report, a rhetorical tactic that distances the reporter even further—is presented for explanatory reasons only. The radical optimism of the EPA's claims about curbing the greenhouse effect, given in the last sentence, is treated with a similar distance.

Ironic Distance

The rhetorical distance cultivated in these reports in the "Research News" and "News and Comment" sections of *Science* occasionally mounts toward an overweening irony. In the conclusion of "Report Urges Greenhouse Action Now," Kerr reflects upon the public consciousness with a grating sarcasm:

> The ozone hole is among the reasons that major environmental groups . . . are starting to put time and money into the problem. But environmentalists will still have their hands full raising the public's consciousness. A recent poll found that two-thirds of Americans believe that the greenhouse effect presents a somewhat to very serious danger. But that placed it thirteenth out of 16 problems, beating out only x-rays, indoor radon, and radiation from microwave ovens. What would be handy is a crisis. No one is willing to call the current drought a greenhouse effect, but it could still become the ozone hole of the movement to control the greenhouse. (24)

In the last sentence, Kerr shows little sympathy for environmentalists who use physical data—the occasional ozone hole or drought—to construct some version of what the cultural critic Walter Truett Anderson has called a "noble lie," a propagandistic use of information to further a worthy cause (qtd. in Prescott 39). The reporter attempts to create irony by distancing himself and his readers in the scientific community from the environmentalist struggle to effect a shift in the public consciousness.

Two nebulously delimited groups—the "environmentalists" and the "public"—are set at odds against one another and are implicitly distinguished from scientific researchers, whose aim presumably has nothing to do with raising the public consciousness. The "no one" of the last sentence might well be translated "no one of us," for the suggestion is that in fact the public has perceived the current drought as an effect of greenhouse warming and that the environmentalists are using that perception to their own political advantage whether or not they can get confirmation from the scientific community. "What would be handy"—from the environmentalist perspective, that is—"is a crisis."

The aim of the irony, then, is to solidify a community perspective—scientific objectivism—by creating an image not of that community itself, but of other communities that are put on display at an ironic distance. By gaining perspective on these communities, the reader comes to think of them as the *other*. Even when the scientific perspective is treated in the third person—as in the sentence "Scientists' views of the future are as murky as ever" (Kerr, "Report Urges Greenhouse Action" 23)—the tone that arises is one of *self-irony*, since scientists, despite their search for "predictive models," tend to associate ordinary predictions of the future (Will it be dry or wet this summer; will the world fry to a crisp by the turn of the century?) with a bygone age of magic and superstition.

In an October 1988 "News and Comment" story, whose very title is sarcastic, if not belittling—"Johnny Appleseed and the Greenhouse"—William Booth writes, "Until very recently, asking how many trees would have to be planted to mitigate the greenhouse effect seemed not only naive, but a bit absurd—the kind of calculation more appropriately presented on a cocktail napkin than before a congressional committee" (19). The clear indication of the implied audience for this sentence is that few nonscientists will have had the experience of doing this kind of calculation on a cocktail napkin. Having thus defined the inner circle to whom the article is addressed, Booth proceeds to make his contribution to the comic portrait of other perspectives. In this case, we are distanced from government agencies as well as environmentalists. The agencies are indulged as careless purveyors of jargon and inexact information produced under political pressure, as the italicized phrases and figures of speech in the following passage suggest:

> *Under the gun* to come up with policy options to control global warming, the Environmental Protection Agency is also taking a serious look at reforestation. "In the long run, *it might be cheaper than a lot of other options,*" says Daniel Lashof of EPA. Lashof adds that *planting trees "provides for a nice synergism,"* since trees not only absorb carbon dioxide and store the carbon as woody biomass, but they also slow soil erosion, improve watersheds, provide timber, and shelter a web of biodiversity. *This kind of laundry list of dividends is the stuff legislation is made of.* A pair of greenhouse bills, one introduced by Senator Timothy Wirth (D-CO) and another by Representative Claudine Schneider (R-RI), both include language on reforestation. (Booth 19; italics added)

The last two sentences extend the irony to the legislative process, portraying bill-making research as a ragtag collection of handy language prefabricated in the government agencies. Environmentalists are also portrayed as a careless lot, hungry for any new angle from which to peddle their political bill of goods: "Environmentalists are wasting no time selling the scheme" (Booth 19). Calculations of how many trees would be needed to "do the trick" are described as "rough" and "admittedly crude" (19), and it is suggested, with deep irony, that the figures may even make "a case against reforestation, since the task seems almost too enormous" (19). The far from surprising conclusion of the article is that "More study on these options is clearly needed" (20). But the author goes even further: "And indeed, without stopping or slowing the deforestation that is consuming millions of hectares of tropical forest every year, talking about reforestation seems *out of touch with reality*" (20; italics added).

In *Science*, the reader is thus assumed to identify with the dominant perspective of the scientific community and to feel distance from such perspectives as those of government agencies, decision-makers, environmentalists, and the general public. By contrast, in the reports on global warming and related issues in *Time*, the overall direction of the rhetorical appeal turns to the proverbial common reader, whose historical suspicion of science, government, and environmentalist activism (or any other kind of activism) has begun to yield in recent years to a general acceptance of what all of these perspectives can contribute to a public version of environmentalist consciousness and action.

Time and Public Environmentalism

Time's overarching commitment to informing the general public and to developing stories with human interest tempers its every report on science, nature, and the environment and leads the magazine in some directions predictably different from those we have noted in *Science*. The timing and number of reports on global warming represent the only nontrivial similarities. For nearly a decade beginning in the late 1970s, *Time* offered a section on "Environment" and turned out stories for that section at the rate of about ten per year. But between July 1988 and July 1989, there was a shift in emphasis: three cover stories appeared on environmental topics, the unprecedented "Planet of the Year" issue was produced in January of 1989, and stories for the sections titled "Environment" and "Nature" became nearly a weekly feature.

Like *Science*, *Time* attempts to enhance the drama of its news stories by drawing out potential and actual conflicts in the arena of environmental politics. The set of conflicts chosen by each of the magazines is a fine index of which perspective prevails in its coverage. *Science* has fairly consistently pursued the conflicts within the scientific community about the interpretation of new data and the conflicts between the scientific researcher's outlook and the image formed by the world at large (the forces of politics, the media, and the general public) of scientific information and action. *Time*, on the other hand, reflects what it perceives to be the dominant view on environmental issues, with a special focus on economic questions and conflicts related to regional, national, and international identities.[3]

Before 1988: Ecospeak and Public Identity in Time

Until the summer of 1988, when the magazine underwent a major shift in tone and outlook, *Time* remained true to the categories of ecospeak, treating the developmentalist perspective as the dominant viewpoint of the American public and treating reform environmentalism as a protest movement and a minority position. In addition, stories in *Time* were inclined to emphasize traditional conflicts involved in sectional and national rivalries. Reading *Time*, we have come to expect discussions about the economics of environmental policy and to witness the finger pointing and assignments of praise and blame that accompany any effort at shaping national and international politics.

Brief treatments of global warming began to appear frequently in the pages of *Time* during 1987. In an October report on the Montreal accord on emission control of ozone-depleting substances, for example, the scientific findings are offered in a highly compressed form along with a short apocalyptic story:

> Scientists estimate that overall as much as 7% of the ozone belt, which stretches six to 30 miles above the earth, has already been destroyed. Moreover, researchers have found evidence of "holes" in the shield, including one above Antarctica that approaches the size of the continental U.S. As the world's ozone layer deteriorates, the sun's radiation could lead to a dramatic increase in skin cancer and cataracts, along with a lowered resistance to infection. It could damage plant life, both directly and as a result of a general warming trend; that warming could lead to a disastrous rise in sea levels. (Garelik 35)

Two aspects of this report are typical of *Time*'s coverage of science in issues before the summer of 1988. The first is that the scientific community is portrayed at a distance as a unified social contingent basically in agreement even in matters that are still warm and in the process of becoming factual: "Scientists estimate," "researchers have found." An August 1987 cover story on the erosion of shorelines—

intended no doubt for the eyes of summer beach-goers—transmits this image of the scientific community: "It may be years before scientists determine just how significant the greenhouse effect is—but they know the process is accelerating. Sea levels are expected to rise at least a foot in just another half-century" (Lemonick, "Shrinking Shores" 44). The second aspect typical of *Time* is this: In reports on scientific findings, the emphasis falls squarely upon the direct effects on human life—the possibility of health problems that could result from ozone depletion—with indirect effects, like damage to plants and rising sea levels, getting a secondary emphasis, and with no mention of theoretical implications or difficulties.

Although placed in the section on "Environment," the story on the Montreal ozone accord is really most deeply concerned with international politics and with economics. The Soviet Union's foot-dragging on signing the pact receives a treatment as full as the presentation of the scientific testimony, as does the Canadian government's raising of the sore issue of an acid rain accord with the United States. In 1989 articles, it comes to be Japan—America's most efficient adversary in international economics—that is cast as the villain in the story of world environmental protection by reporters in both *Time* (Linden) and *Newsweek* (Begley). Thus the treatment of international environmental politics often reflects the allegiances, suspicions, and tensions that generally beset American military and economic relations with other countries.

Economics certainly are at the center of all discussions. A full paragraph in the ozone story is devoted to potential costs of limitations placed upon the production of ozone-depleting chemicals, with a special place reserved for the response of one multinational corporation to the proceedings: "The costs of the treaty could prove considerable. CFCs have become popular because they are generally safe to apply and relatively cheap to produce. But Joseph Steed, environmental manager for DuPont, whose annual production of CFCs (under the brand name Freon) is valued at $600 million, warns that adoption of the protocol will mean a lengthy and expensive search for alternatives and that the costs will be passed on to consumers" (Garelik 35). The rhetoric of this passage, in an appeal to the cost-conscious consumer who makes up the "general public" to whom the magazine is primarily directed, is strongly conservative. We are told that CFCs are "popular," "relatively cheap," and—in a puzzling phrase, considering that we have just heard about the potential health hazards of ozone depletion—"generally safe to apply." The short-term concern with cost, we must guess, is paralleled by the fact that CFCs pose little immediate threat in the workplace and hence are "safe to apply."

This conservative tone and this implicit argument based on short-term costs—like the skeptical caution that characterizes the reports in *Science*—represents an effort to predict and cater to the most likely response of the target audience. It reflects what, up through 1987, the authors at *Time* could presume was the dominant attitude of the American public about environmental issues—a mistrust of sweeping changes based on the fear that, whatever the actions of high-minded scientists, reform environmentalists, and international politicians, the greatest costs will be borne by the ordinary citizen.

Unless there was an immediate connection with economics, a story on the environment had little chance even of being included in *Time* before 1988. From the October 1987 issue that contains the report on the Montreal accord until the summer of 1988, we find stories in the "Environment" section, none of them over three columns in length, on such topics as the return of life following the eruption of Mt. St. Helens (boding well for the tourist industry), the banning of a termite pesticide, an oil spill, the role of acid rain in the destruction of spawning grounds of ocean fish, conservationists' battle with the users of all-terrain vehicles on public lands—all treating the conflict of environmentalists and developmentalists in varying degrees of emphasis. Except for a short sidebar in the issue of 30 May 1988 on the ozone hole, nothing more appears on global warming and related issues, not even in the first major report that *Time* prints on the drought in the 27 June number.

The Year 1988: The Shift toward Environmental Awareness

The changes begin on 4 July 1988. In the issue of that date, major environmental questions are raised in the cover story on the U.S. drought, which is placed not in the "Environment" section but in the "Nation" section. In addition to the featured report on the extensiveness of the drought—"The Big Dry" (Sidney)—a full page is devoted to a related article on global warming. The title asks the question, "Is the Earth Warming Up?" and the subtitle answers, in a scientifically honest anticlimax: "Yes, say scientists, but that may not explain this year's heat wave" (Brand 18). Nevertheless, the article begins with a prototypical apocalyptic story in italics, followed by a condensed explanation of the greenhouse theory and a generously full treatment of Hansen's testimony before the congressional committee. The result is high journalistic drama:

> *The Great Plains has become a dust bowl, and people are moving north into Canada's uplands to seek work. Even in Alaska, changing ocean currents are boosting the fish catch. New York is sweltering in 95° weather that began in June and will continue through Labor Day. In the Southeast the hot spell started six weeks earlier . . .*
>
> That picture of the future is all too familiar to many meteorologists. To some, it makes the drought that is crippling the nation's midsection seem an ominous harbinger of things to come. Because of the greenhouse effect, a process by which natural and man-made gases trap solar heat in the earth's atmosphere, the gradual warming of the globe is inevitable, in the view of many scientists. But until now, most had cautiously avoided definitive statements about precisely when such an effect might take place.
>
> Testifying before a congressional committee last week, James Hansen, an atmospheric scientist who heads NASA's Goddard Institute, riveted Senators with the news that the greenhouse effect has already begun. During the first five months of 1988, he said, average worldwide temperatures were the highest in the 130 years that records have been kept. Moreover, Hansen continued, he is 99% certain that the higher temperatures are not just a natural phenomenon but the result of a build-up of carbon dioxide (CO_2) and other gases from man-made sources, mainly pollution from power plants and automobiles. Said Hansen: "It is time to stop waffling and say that the evidence is pretty strong that the greenhouse effect is here." (Brand 18)

The implication that Hansen, rather than being a free-lancer and renegade, is a major spokesperson for the scientific community at large—one of the "many scientists" who "until now . . . had cautiously avoided definitive statements"—is tempered with quotations from Stephen Schneider ("It doesn't prove the greenhouse effect") and

Chester Ropelewski ("The hard evidence isn't there"), but these appear late in the article well after the dramatic presentation of the apocalyptic vision of the future that is said to be "all too familiar to many meteorologists" and the portrait of the prophet Hansen "riveting" the senators. The rhetorical effect achieved in the first three paragraphs could hardly be dampened by the later inclusion of the standard scientific line of caution. If anything, these obligatory demurrals enhance the drama of Hansen's refusal to "waffle."

This key issue of 4 July appears to signal the conversion of *Time* magazine to the environmentalist cause. From this point onward, *Time* devotes an unprecedented number of columns to ecological questions, within a month presenting another cover story on a major environmental issue—the pollution of the oceans (Toufexis, "Dirty Seas," 1 Aug. 1988)—as well as expanding stories in the "Environment" section to two or three pages each, and even covering matters of ecological theory—such as the La Niña phenomenon—in the "Science" section (Linden, "Big Chill for the Greenhouse," 31 Oct. 1988).

Moreover, the slant of the treatment of environmental topics undergoes an interesting rhetorical alteration. Though still committed to dealing with economics as a primary concern, the reporters frequently extend their analysis to sociological and historical matters. A 5 September story on the landfill problem predictably reports on the swelling garbage budget of most U.S. towns but also suggests apocalyptically that the day is hastening when "there will be no place to dump garbage" (Church 81) and even suggests that "in large part, the garbage crisis is a cultural crisis" (82). "Cultural change is notoriously slow," the article instructs, "but it might be speeded up in this instance by the lash of crisis" (82).

Borrowing this metaphor, we might suggest that it is the "lash of crisis"—a crisis made to seem ever more real by the authority of Hansen's testimony—that has driven the conversion of *Time* to the environmentalist perspective. We may wonder, as we would with all converts, about the depth and sincerity of this new direction and may remember the outpouring of reports on alternative energy sources during the "energy crisis" in the late 1970s and the consequent abandonment of interest in solar, geophysical, and wind energy when relatively cheap oil began to flow again. Slight backsliding on the global warming issue is already apparent in the 6 February 1989, report "The Forecast: Hazy and Puzzling," subtitled thus: "A study says the U.S. is still cool, but the greenhouse theory lives." Though placed on the defensive by the reporter in this story, Hansen nevertheless retains his position as a major character in the drama and generally emerges triumphant, "not at all swayed by the new study" of data gathered at weather stations between 1985 and 1987 by meteorologists at the National Oceanic and Atmospheric Administration. Hansen is quoted as claiming that "you'd still expect on a statistical basis to see local variation. . . . But it's the global average that is important." The reporter is apparently convinced, as he awards Hansen the last word in the article, which ends with this quotation: "Our model predicts that by the middle of the 1990s, the greenhouse effect should be pretty clear not only to scientists, but also to the man in the street" ("Forecast" 57). No doubt "the greenhouse theory lives," and so does its prophet.

On the whole in early 1989, the conversion of *Time* to environmentalism has remained impressive, nowhere more so than in the "Planet of the Year" issue of 2 January 1989. The human interest slant of the magazine, under the lashing of economic fear and necessity, has been extended to a "whole earth interest." The publishers and editorial staff appear to have looked beyond the view of nature-as-resource toward the view of nature-as-spirit (or nature-as-being). In 1982, *Time* had featured the computer as the "Machine of the Year," indicating a trend toward an objectification or mechanization of its primarily human-centered perspective. But by 1989, nature had caught the magazine's interest with a crisis-level insistence. As in the mythic personification of the earth as the goddess Gaia, *Time* featured its home planet in a position normally reserved for human subjects. The personification of the planet represents a significant step toward the rights-of-nature approach of the deep ecologists. The publisher Robert L. Miller offers this explanation of the issue, which features on the cover a globe wrapped in polyethylene and rag rope (in a design invented by the Bulgarian-born environmental sculptor Christo): "This week's unorthodox choice of Endangered Earth as Planet of the Year, in lieu of the usual Man or Woman of the Year, had its origin in the scorching summer of 1988, when environmental disasters—

droughts, floods, forest fires, polluted beaches—dominated the news. By August *Time* knew it was no longer enough to describe familiar problems one more time" (R. Miller 3). In November of 1988, Miller explains, *Time* invited a team of experts to a conference for an interchange with its staff, and the process of education and dialogue that ensued stimulated the bulk of the copy for the special issue and set the agenda for an entire year of new emphasis on the environment.

A New Focus on Action

Not only does the special issue set out to educate the general public on the prime issues of environmental degradation—the need for biodiversity, the problem of toxic wastes, the destruction of the rain forests, and so on—but also provides after each article an international action agenda. Indeed, the focus of the whole "Planet of the Year" issue is on action. This emphasis suggests that like the science textbooks we have studied but unlike the science news journals (except in editorials and essays by individuals—Koshland and Schneider in *Science*, for example), *Time* is doing more than merely reporting the facts; it has taken up overt efforts to influence future actions.

In the special issue story on the greenhouse effect, Lemonick's "Feeling the Heat," for example, the overall faith in the greenhouse theory is asserted strongly in large bold print in the article's subtitle: "Greenhouse gases could create a climatic calamity." Interestingly enough, the extreme pronouncements of Hansen's position are brought under harder scrutiny than in any other *Time* article. "Many climatologists call Hansen's remarks premature and feared that if this summer happens to be cool, public worries about the greenhouse effect will quickly fade," the author writes, attaining for the moment at least a low pinnacle of ironic distance. "But," the article continues, drawing upon the "global experiment" metaphor that scientists have used since the late 1950s, "no one disputes the fact that the amount of CO_2 in the atmosphere has risen and continues to increase rapidly and that the human race is thus conducting a dangerous experiment on an unprecedented scale."

Having drawn a fairly careful distinction between controversial ("warm") issues and agreed-upon ("cold") facts in the scientific discussion, the reporter turns up the rhetoric: "The possible consequences are so scary that it is only prudent for governments to slow the buildup of CO_2 through preventative measures, from encouraging energy conservation to developing alternatives to fossil fuels" (Lemonick, "Feeling the Heat" 36–37). The action alternatives are developed in some detail in the rest of the article and for emphasis, are summarized in a sidebar on the second page of the report under the heading "What Nations Should Do":

> 1. Impose special taxes on carbon-dioxide emissions, which would encourage energy conservation.
>
> 2. Increase funding for research on alternative energy sources, including solar power, and safer designs for nuclear reactors.
>
> 3. Provide financial aid to enable developing nations to build high-efficiency power plants rather than conventional facilities.
>
> 4. Launch a mammoth international tree-planting program.
>
> 5. Develop techniques for recovering part of the methane that is given off by landfills and cattle feedlots. (Lemonick 37)

Several features of the liberal environmentalist perspective are worth noting here: the insistence upon large-scale intervention by big government; the faith in scientific research to provide crucial answers; and a corresponding faith that technological solutions can be found to technological problems. What had seemed a controversial position for Rachel Carson and Barry Commoner in the 1960s and 1970s thus reached a high point of public acceptance by the late 1980s, a level of acceptance symbolized by *Time*'s willingness to embrace the values of public environmentalism. Other well-known publications—among them *Newsweek*, *Scientific American*, and the New York *Times*—have followed similar paths.

Is the News Sustainable?

Time and other general news magazines have discovered a central premise of ecological humanism: Human interest resides in the preservation of nature. The media now produce a continuing discourse that significantly shapes the public character of environmental consciousness. Despite its broad appeal and its potential influence, however, this discourse cannot carry the full narrative function of the growing environmentalist movement of American culture. For one thing, the field of politics is wide, and the news media must sacrifice deep coverage for breadth of coverage. Another problem is that the news is conventionally tied to recent events rather than to ideas and historical and future conditions. *Time*'s move into the field of policy in its special issue on the endangered earth was justified by the extreme events of 1988 and by the retrospective and hence reflective character of the "Planet of the Year" issue. The development of an environmentalist action agenda is not a sustainable direction for either scientific research (which has its own independent action agenda), or science journalism (with its strong links to the scientific community), or finally the mass media (with its interest in reporting the news).

In the search for discourse types that fill this gap and take up instrumental and rhetorical writing in the hope of bringing consciousness to bear on action, we turn in the next three chapters to government policy documents (chapter 5), environmentalist fiction and other forms of symbolic action (chapter 6), and the discourse of economic advisors (chapter 7).

No one knows who will live in this cage in the future, or whether at the end of this tremendous development entirely new prophets will arise, or there will be a great rebirth of old ideas and ideals, or, if neither, mechanized petrification, embellished with a sort of convulsive self-importance. For of the last stage of this critical development, it might well be truly said: "Specialists without spirit, sensualists without heart; this nullity imagines that it has attained a level of civilization never before attained."

—Max Weber,
The Protestant Ethic (182)

[The] vision of new possibilities requires only the recognition that scientific discoveries can be used in at least two opposite ways. The first leads to specialization of function, instrumentation of values, and centralization of power and turns people into the accessories of bureaucracies or machines. The second enlarges the range of each person's competence, control, and initiative, limited only by other individuals' claims to an equal range of power and freedom.

—Ivan Illich,
Tools for Conviviality (xii)

5

The Environmental Impact Statement and the Rhetoric of Democracy

Instrumental Rationality in Government Resource Management

The historian Samuel Hays argues that a division has evolved in American environmental politics between, on the one hand, *experts* who have institutionalized access to authoritative information and influence, and on the other hand, the *general public*, whose sources of information and whose power to influence policy remains uncertain. According to Hays, the expert "thinks of the political context as one of 'us' and 'them,' of the knowledgeable and rational experts and the uninformed and emotional public" (*Beauty* 9). This division generates a major rhetorical rift in public writing on the environment, resulting in competing discourses, the contrary aims of which can be discussed in terms introduced by the German philosopher and social theorist Jürgen Habermas—"instrumental action" versus "communicative action."

Whereas the goal of public participation in environmentalism is tied to the need for communicative action, the expert conservationists who hold the reins of regulatory action in government have, since the turn of the century, sought to instrumentalize the relationship of people and resources. The rhetoric of instrumentalism dominates the major documents produced by administrative government, including the controversial and much-discussed genre known as the environmental impact statement.

Who Are the Experts?

The experts whom Hays identifies as applied scientists and technicians are not usually a part of the mainstream scientific community that we have been concerned with in the last three chapters. They are not academic or research scientists committed to the theoretical project of basic research or even to the goal of sustaining scientific inquiry. Their primary objectives are directed instead to sustaining governmental control and perpetuating a rational social order. They are interested in facts, not theory, and in procedures, not scientific or political action per se. On the map of discourses we offered in figure 1 of our introduction (see page 11), these experts occupy the subject position of "government."

The government experts require huge compilations of information, from which base they are able to assert their own authority (ethos). They have preserved the scientific interest in inductive reasoning and in the generation of data, but have otherwise ignored the procedures of scientific learning, especially the rules of tight argumentation (Latour) and the rules for the social construction of knowledge—the need for peer review and careful comparisons of data with research standards developed from the work of other researchers.

Whereas the discourse of research scientists springs from a long tradition of inquiry, with roots reaching back to the seventeenth century and ultimately to classical philosophy, the experts who occupy the government bureaus rose to prominence only within the last one hundred years. Their approach to the management of people, resources, and information originated in the Progressive Era, when Frederick Taylor, the first "efficiency expert," introduced "scientific management" into American industry (Noble 264–74). Just as Taylorism demonstrated industrial management's lack of confidence in labor and denial of the worker's contribution to creative aspects of production, the advent of scientific management in American government was accompanied by a devaluation of the potential contribution of local governments and the general public to national policy. Research scientists may share with scientific managers a mistrust of the public, but they also look upon these government experts with profound suspicion (see Schneider, *Global Warming* 204).

However we arrange the categories of *us* and *them*, though, the public emerges in the position of *them*, the ones who lack sufficient information and influence, thus leading to Hays' compelling interpretation of the development of conservation politics in America:

> The first American conservation movement [1890–1920] experimented with the application of the new technology [of Taylorism or scientific management] to resource management. Requiring centralized and coordinated decisions, . . . this procedure conflicted with American political institutions which drew their vitality from filling local needs. This conflict between the centralizing tendencies of effective economic organization and the decentralizing forces inherent in a multitude of geographical interests presented problems to even the wisest statesman. (Hays, *Conservation* 275)

Whereas environmentalism has become a broad-based public movement arising from the history of material *consumption* in American life, conservationism has its roots in expert control of material *production*. For Gifford Pinchot, the founder of both the U.S. Forest Service and the science of forestry, as well as for his friend and boss, President Theodore Roosevelt, "Conservation, above all, was a scientific movement" whose "essence was rational planning to promote efficient development and use of all natural resources" (Hays, *Conservation* 2). Roosevelt argued that "the conservation of natural resources . . . is yet but part of another and greater problem . . . the problem of national efficiency, the patriotic duty of insuring the safety and continuance of the Nation" (qtd. in Hays, *Conservation* 125). The Roosevelt administration consistently sought the advice of a professional corps of scientifically trained experts, for whom Pinchot himself provided the model. According to the outlook of the new resource specialists, the public was a force to be neutralized, not incorporated into the decision-making process.

Instrumental vs. Communicative Action

In our own day, as in Pinchot's day, the resource specialists in the government bureaus operate according to the patterns that social theorists from Max Weber (a contemporary of Pinchot and Frederick Taylor) to Jürgen Habermas have called *instrumental rationality*.

The agencies' resistance to involving the public in policy decisions precludes the operation of *communicative rationality,* which Habermas described with great thoroughness in his theoretical works of the 1980s.[1]

In the words of the American social theorist Mark Poster, "Instrumental rationality [as conceived by Habermas] characterizes practices in what he calls 'the system,' that is in institutions like the bureaucratic state and the economy, which achieve social solidarity through 'steering mechanisms' " (Poster 23). Documents motivated by instrumental rationality have as their sole purpose the control of the document's readers. These writings may take on the *appearance* of traditional scientific papers, whose purpose is to persuade readers to accept an interpretation, usually amounting to a change of direction in a research program. But instrumental documents are not really interested in interpretation or in persuasion; they attempt to create, for the purpose of maintaining the system, a narrow path of action that has been chosen or created in advance of the document's production by hierarchically arranged powers. And though they may draw upon the conventions of a democratic discourse that is open to information from diverse sources, the aim of instrumental documents is never to treat deviant discourses with respect but always merely to take note of them, to record them, and ultimately to treat them as "noise" in the system, which needs to be ignored or expunged.

Communicative rationality, on the other hand, "characterizes actions in what [Habermas] calls the lifeworld, that is, in areas of social action where socialization and cultural reproduction are at issue" (Poster 23). According to Habermas, "*communicative rationality* carries with it connotations based ultimately on the central experience of the unconstrained, unifying, consensus-bringing force of argumentative speech, in which different participants overcome their merely subjective views and, owing to the mutuality of rationally motivated conviction, assure themselves of both the unity of the objective world and the intersubjectivity of their lifeworld" (*Theory of Communicative Action* 1.10). Whereas the systems of instrumental rationality are construed as a hierarchy, with experts steering subjects along designated paths—the great dream of Taylor the efficiency expert, the great nightmare of Orwell the libertarian—the lifeworlds of the subjects of communicative action are construed as the interlocking nodes of a network or the overlapping bubbles of a very complex Venn diagram. Where the lifeworlds of social subjects overlap—encouraged by shared belief systems, historical events, good arguments, and other means of cultural identification—a space for action supported by "intersubjective" consensus is created. Since Habermas' theory of communicative action is rooted in classical Greek political theory (as strained through the phenomenology and praxis philosophy of Hegelian, Marxian, and Weberian social theory, Anglo-American speech-act theory, and American pragmatism), it is no wonder that documents conceived in the mold of communicative rationality would bear the marks of classical rhetoric, especially as portrayed by Aristotle, the first great exponent of praxis philosophy or "practical reasoning."[2]

In technological societies, the liberal love of democracy, with its preference for communicative action, is corrupted by the liberal desire for efficiency, the great goal of instrumental action. The situation is rendered almost desperate in social situations where cultural diversity is great. Communicative rationality insists that, out of this Babel of perspectives, a reasonable course of action will emerge. But instrumental rationality insists that the costs of communicative action in time and money are too high, that people are confused about their own real needs, that impulses and emotions override rationality in public debate, and that good action depends upon expert guidance. Thus it happens that the system is allowed to intrude upon the lifeworld, with the result that communicative concerns over mutual understanding and consensus are replaced by "hierarchically distorted verbal exchanges in which each party instrumentally manipulates the other, with the state, for example, having a considerable advantage in the manipulation game over a welfare mother" (Poster 23). In the concept of instrumental action, as Habermas suggests, "the rationality of self-regulating systems" creates "imperatives [that] override the consciousness of the members integrated into them" and thus "appears in the shape of a totalized purposive rationality" (*Theory of Communicative Action* 2.333). The role of the *citizen* is restructured so that the citizen becomes a *client* of the system. No doubt, the client's material needs are met by government and his or her rights are ensured, but at a cost: "The establishment of basic political rights in the framework of mass democracy means, on the one hand, a universalization of the role of citizen and, on the other hand, a segmenting of this role from the decision-making process, a cleansing of political participation from any participatory content" (*Theory of Communicative Action* 2.350).

As Hays has shown, Pinchot and other conservationists in the government bureaus inaugurated a century-long history of instrumental action, the agencies vying for control with Congress, the general public, and even the president through the process of limiting access to legitimate information and the means to transform that information efficiently into action. As a justification for this claim to power, Pinchot and his descendants have asserted the instrumental superiority of the agencies. They could, according to this rationale, work fast where Congress could only plod. They could be objective when the general public could only burn with emotion. They could accomplish real, useful actions while academic scientists were amusing themselves by contemplating theory. They could promote the general interests of the whole system of government against the favoritism and fickleness of elected officials and the ephemeral moods of a mass public under the influence of distorted information spread by a news media with little or no understanding of the issues reported. As we shall see, their writing asserts their privileged status as experts and is calculated to achieve efficient and systematic action while minimizing such "noise in the system" as public commentary on proposed actions.

The government agencies—the action branch of government, the end point of so many rhetorical appeals—may be more efficient in achieving their self-defined goals, but they are slow to realize and assimilate mass democratic movements like environmentalism. The by-product of their commitment to rigid procedure is an equally rigid method of gathering and reporting information in recommendations for action. Like scientific researchers, they create a discourse that is all but impenetrable for readers outside their own discourse community. Unlike scientific researchers, however, they have fewer mechanisms for adapting to change and for accommodating "an objectivity and impartiality secured through unrestricted discussion" (*Theory of Communicative Action* 2.91). As a consequence, they lag behind in developing a public discourse that meets the public's demand for change.

A rhetorical analysis of one of the leading vehicles of instrumental action in the governmental bureaus, the environmental impact statement (EIS), reveals the cultural inadequacy of strictly instrumental discourse as well as the transparency of the effort to clothe that discourse in the garb of democratic, communicative discourse while maintaining a commitment to the purest forms of instrumental rationality.

Instrumentalism and Style in the EIS

The environmental impact statement arose from the deepening environmental consciousness of the 1960s as embodied in the legislative action of the National Environmental Policy Act (NEPA) of 1969. In the two decades since NEPA passed into law, the number of EISs and the process by which they are produced have grown rapidly, filling volume after volume and packing the archives of government agencies. The EIS has now become the primary type of discourse connected with actions involving public lands in America. Spawned in the conflicts between the legislative and executive branches of American government—the same kinds of conflicts documented in Hays' study of conservationism in the Progressive Era—the EIS has remained controversial throughout its short history. As far as we can determine, it is unprecedented as a governmental effort to control environmentally related practices indirectly by controlling the discourse associated with those practices.

Open in Theory, Closed in Practice

During the drafting of NEPA, EISs "were invented in response to the anticipated administrative indifference or outright hostility toward" the new legislation (Dreyfus and Ingram 251). In its completed form, the law requires federal agencies, and in some cases private industries, to issue an EIS before effecting or allowing a change in areas of land, water, or air that fall under federal jurisdiction. The EIS is supposed to describe a proposed action, the reasons for it, and any short- or long-term effects. Using the same methods of analysis, it must also consider optional plans for managing the environment in question.

The Congress saw fit to require the agencies not only to carry out environmental studies and to prove that they had done so by writing these "action-forcing documents," but also to open the process for public comment. The original legislation, therefore, fitted out the EIS as a vehicle of communicative action. But it made no provisions for restructuring or re-funding the agencies involved so that this possibility could be realized.

Instead, the power of expertise and the gospel of efficiency prevailed. The principal authors of the EISs are usually government agents with scientific training. Occasionally hired consultants will supplement the findings of these agency employees, but the bulk of the work is done by in-house "Resource Specialists," as they are called in the parlance of the Department of the Interior. These agents are assisted by government technical writers and editors. The research and writing are nearly always a team effort, but little or no effort is made to include contributions from commentators outside the inner circle of the authoring agency. The primary readers of the EIS are people who make decisions about land use and air and water quality—executive administrators and sometimes judges and legislators. The intended audience also consists of invited commentators, related government agencies, and concerned citizens—all of whom may in principle influence the final decision of the primary audience through testimony, advice, lobbying, and voting. Our research, however, like that reported in other studies of EIS, shows that the likelihood of an outsider influencing an agency action is slight. While the system constructed, maintained, and reproduced by the EIS process has little or no effect upon the lifeworld of the agency experts and primary decision-makers, those whose worlds are most deeply affected are systematically excluded from participation in the process, even while their rights to be heard are ostensibly maintained. The very language of the EIS ensures this exclusion of the interested public. The authors strictly maintain the rhetorical conventions of the "objective style" in their presentation, thus manifesting the ethos of detachment associated with scientific investigation. This stance further closes the communicative discourse process by hampering the general readability of the EIS; it makes the information of the EIS least accessible to those who want it most—decision-makers and politicians outside the agency and the people who seek to influence them (and whose interests the decision-makers ostensibly represent).

The Expert's Rhetorical Objectivism: A *Case*

Our analysis of EIS rhetoric is based on a 1985–86 study, in which (with the help of Dean Steffens) we subjected to rhetorical criticism a library of EISs developed for projects in central New Mexico and filed in the Socorro District Office of the U.S. Bureau of Land Management (BLM). In addition, we analyzed the U.S. Department of the Interior *Editorial Management Handbook for Environmental Impact Statement and Environmental* Assessment (1980), and we interviewed a number of writers and resource managers working in the Socorro office. In 1988, we followed up the initial study with telephone interviews. Our aim was to get at the particular traits of the government expert's rhetoric as a step toward evaluating the effectiveness of the EIS as a communicative document.

The predominant tone in the BLM documents is one of disinterest. The perspective requires distance from subject matter and audience. Despite any personal feelings the various authors may have—several we interviewed expressed a deep love of the region they studied and a commitment to the New Mexican people as well as to environmentalist values—the language of the expert nullifies potential identifications with the ordinary reader and with the environment that is examined, both the "physical environment" (land, air, water, plants, animals) and the "social environment" (people). The objective tone does, however, preserve a social identification within the community of experts; the very identity of the expert depends on this language of disinterest and distance. As surely as it objectifies the land, water, air, plants, animals, and people that constitute the environment under study, this objectification extends even to the self—the personality of the author is expunged by impersonal constructions and passive voice—and to those of the author's immediate social group (other technical experts reduced to names attached to studies that compose the technical "literature"). This implicit denial of the self attempts to replace a subject-subject relation (between the authors and their readers) and a subject-object relation (between the authors and the environment in question) with an ideal object-object relation between the expert and the objects of study. The EIS author presents indirectly a self-portrait of a data-gathering machine, a computer with an optical scanner and word-processing capabilities.[3]

No contempt for the ordinary reader is necessarily intended in this approach. On the contrary, in the EISs we examined, there is good evidence of the author's efforts to follow the injunction on terminology given in the *Editorial Management Handbook*: "The neçessity of maintaining the readability of an EIS . . . requires that the Resource Specialists must carefully ensure that only that level of technical language essential to the average reader's understanding is retained" (7–3). To meet this goal of communication, the authors courteously define technical terms, at least on first use, and provide glossaries of important words. Of course, saturation may still result from the introduction of too many technical terms in a short space. But for the most part, familiar words seem to be preferred to technical terminology. Native plants, for example, are called "saltbush" and "mesquite" rather than by some Latin words unfamiliar to most farmers and ranchers, and to avoid confusion (how many desert plants are called "saltbush"?), photographs are included.

The approach to language of the *Editorial Management Handbook* is nevertheless a slight and ultimately trivial step toward effective communication. It is weak because it is atomistic; it keeps the authors' and editors' attention focused on individual words. The treatment of single terms may demonstrate an honest concern with the public readership and may decrease slightly the characteristic detachment of scientific writing, but the detachment and the general tone of cool efficiency are thoroughly preserved in the syntactic and structural features of the EISs. Not the individual words themselves but the combinations of them place this prose into the category that Richard Lanham has named "voiceless." The authors prefer the "noun style," which favors expressions of stasis ("Arrival; Reconnaissance; Victory"), over the "verb style," which favors expressions of actions ("I came. I saw. I conquered.") (Lanham 15). The verb style pictures a world full of human actors performing purposeful actions upon objects in an ever-changing scene; it requires active verbs and human subjects, as well as a full range of adjectives, adverbs, phrases, and clauses that delineate subtleties of modification and relation. By contrast, the objectivist syntax of the noun style—expressive of a world frozen into stasis and broken (analyzed) into its odd components—is dominated by features like passive voice, nominalizations, strings of noun modifiers, grammatical indefiniteness, impersonality, and high levels of abstraction.

The features of objectivist style have been diagnosed in bureaucratic and technical prose and have been treated with editorial antidotes elsewhere (Williams; Killingsworth, "Thingishness"). Here we introduce them not so much as objects of revision but as the chief syntactical means by which the authors of EISs achieve distance from their subject matter and audiences. Following are some features of objectivist style that we discovered in our reading of the EISs:

Passive Voice *(obliterates agents of actions and thereby obscures responsibility and/or authority):*

Sample 1: Two herbicides, Grasland and Tordon 10K, are currently proposed for use on pinyon-juniper and rabbitbrush encroachment. . . . both are pelletized, and would be broadcasted aerially or by hand on each individual plant species. . . . (U.S. BLM, *Draft Environmental Impact Statement* 1-19)

Sample 2: Prescribed burning on 234,880 acres would be conducted only during periods that would disperse smoke, thereby causing only very short duration, minimal impacts on air quality. (U.S. BLM, *Draft Environmental Impact Statement* 3-3)

Sample 3: The Forest has been inventoried for visual quality. (U.S. Forest Service , 117)

Nominalizations *(favor stasis over action by using words in their noun forms when they might just as well be written as verbs or adjectives):*

Sample 1: Improvement in the *naturalness* of the areas would occur as a result of eliminating or curtailing vehicle *use* on 153 miles of vehicle ways. (U.S. BLM, *New Mexico Statewide Wilderness Study* 1; italics added)

Sample 2: VQLs [Visual Quality Levels] of *preservation, retention, partial retention, modification* and maximum *modification* are assigned to each based on the inventory criteria. The criteria include

> *visibility*, number of *viewers*, and the *uniqueness* or *variety* of a landscape. (U.S. Forest Service 117; italics added)
>
> ***Strings of Noun Modifiers*** *(obscure relationships among people and things by increasing the number of nouns and extending their function to replace that of adjectives, adverbs, prepositional phrases, and dependent clauses):*
>
> *Sample 1:* ESRVA = Enhancement of Sensitive Resource Values Alternatives. (U. S. BLM, *East Socorro Grazing* 2-89)
>
> *Sample 2:* Dispersed *recreation capacity* was determined using the *Recreation Opportunity Spectrum (ROS) analysis* conducted during the development of the Analysis of the *Management Situation* (AMS). Capacity, including *wildlife recreation* but excluding wilderness, is 2,226,000 *recreation visitor days* (RVD's) annually. (U.S. Forest Service 110; italics added)

No doubt, general readability and comprehension suffer because of such writing. More importantly, the expert's style limits access to the information of the EIS to those accustomed to reading and interpreting this form of discourse.

The Costs and Benefits of Stylistic Economy

Another point of style suggesting the discourse of a closed community is the trend toward extreme condensation of expression—the proliferating acronyms in the BLM's EISs, for example. The problem with amassing acronyms is that not only do they cause the reader unused to government jargon the inconvenience of constant page turning to locate original references or lists of abbreviations, but they also increase the density of information that must be absorbed in a short space. Largely because of this density, most EISs intimidate the average reader. The authors of the "Standards of Style and Usage" section of the *Editorial Management Handbook*, perhaps unwittingly, encourage such style. "The economical use of words is stressed in the new CEQ Regulations," the manual reads, without indicating in the immediate context what the CEQ is, "and every review looks for ways to shorten the text without sacrificing clarity and meaning" (U.S. BLM, *Editorial Management Handbook* 7–2). This passage, very typically, gives advice without examples or methods by which to achieve the goals it sets. Though it does not give examples, however, it certainly sets them in its own use of acronyms.

In addition to risking saturation of the reader's capacity for absorbing information by using great numbers of acronyms, the authors also favor high-density graphics like long, multicolumned tables with nested categories and multiple footnotes, elaborate technical mapping, and detailed photographs. High-density graphics are powerful summarizing tools, highly efficient in their ability to display large amounts of comparative data in very little space. They are, for these reasons, potentially valuable instruments for decision making (Bertin). But several commentators on graphics (Tufte, for example) warn that such graphics, like the acronyms, can lead quickly to information overload.[4]

The systems of management by which EISs are produced seem to prefer high-density efficiency or stylistic "economy" to rhetorical effectiveness. This preference is quite explicit in the "Standards" section of the manual: "Every word and every revision in an EIS, in effect, costs money, and this should be the prime consideration in EIS writing" (U.S. BLM, *Editorial Management Handbook* 7–2). The EISs therefore compile information economically without communicating it effectively. An ordinary reader is quickly saturated and easily frustrated by the resulting prose.

Relating to the Audience

The problem of effectively relating to the audience is certainly not ignored in the "Standards," for even the most cynical view of this discourse requires that it maintain a semblance of interest in the needs of the ordinary citizen. Under the heading "Clarity of Text" appears this advice: "The EIS is primarily intended for the decisionmaker and the decision-making process; however it must also be written in a manner conducive to public understanding" (U. S. BLM, *Editorial Management Handbook* 7-2–7-3). This seems to follow fairly closely the general advice on the style of feasibility studies given by Houp and Pearsall in their popular textbook on technical writing:

> Generally speaking, the users are not experts in the field of the study. The users of an environmental impact statement may be citizen

> groups and state legislators. . . . As the writer of [this kind of] feasibility study, you should write in plain language, avoiding technical jargon when possible. Give necessary definitions and background information. Use suitable graphics. Emphasize consequences and function over methodology and theory. Interpret your data and state clearly the conclusions and recommendations that your best professional judgment leads you to. (Houp and Pearsall 384–385)

But note two important differences in the sets of advice. First, the "Standards" introduce a dichotomy of primary audience (decision-makers—upper-level agency officials) and secondary audience (judges, legislators, citizen groups, and private individuals) and thus imply that one level of stylistic sophistication or difficulty is appropriate for the decision-maker while quite another one is needed for "public understanding." This distinction corresponds to the division of us (experts inside an established community) and them (the general public). Houp and Pearsall seem to have a different set of decision-makers in mind—those who lack technical expertise and whose work lies outside the boundaries of the government agency. They thus suggest the need for stylistic simplicity for all potential audiences of the document.

Second, the "Standards" fail, once again, to suggest a method for achieving the stylistic goals set by their advice, whereas Houp and Pearsall point to definite styles and structures. In this way, the "Standards" contribute to the exclusion of interlopers in their community. The only way to learn the style of the EIS is to work within the community. No outsider is allowed access to the esoteric rules of discourse production through this public and widely available style guide.

The trend toward obfuscation of actual guidelines and toward keeping the actual rules of writing within the community is even clearer in this passage from the BLM "Standards":

> Generally, existing guidelines . . . suggest that an EIS be written in a style and terminology understandable to "the average newspaper reader." This is somewhat misleading, as it is not necessarily intended to imply writing in the journalistic style. In its newspaper context, the journalistic style does not have the same intent or goals as the EIS "style" of writing. (This is not suggesting journalism is

> "bad" writing, but rather that the style is not particularly suitable to EIS work, nor is writing for a newspaper-oriented audience.) This Handbook suggests that the best writing is that which expresses the meaning in as few words as possible, and that this "style" will be understandable to any of the Bureau's publics. (U. S. BLM, *Editorial Management Handbook* 7-2–7-3).

We can only guess about the placing of the quotation marks on the word *style* here. Do the authors indicate a recommended avoidance of style in the journalistic or literary sense? Or do they betray an awareness that a simple injunction to brevity hardly constitutes a style?

The "Standards" do advise, as we mentioned earlier, that "Resource Specialists must carefully review their written material to ensure that only that level of technical language essential to the average reader's understanding of the subject is retained" (U.S. BLM *Editorial Management Handbook* 7-3). This suggestion requires a great deal of interpretation, especially the key terms "technical language," "essential," and "average reader." Again, no examples are given, so this advice joins the other goals such as "clarity, simplicity, objectivity, and consistency" at a very high level of abstraction that is of little use to the practical technical editor and "Resource Specialist" in the BLM regional office. As Joseph Williams writes, "We don't lack words to praise good writing: clear, direct, readable, precise. But words like these reflect only how we feel about writing: they don't tell us what good writing is" (8).

Forays in Democratic Discourse

Much of our critique of the BLM documents offered so far is rather too easily reducible to editorial quibbles. Too great an emphasis on this approach would tend to bog us down in details and would keep us from realizing that merely to substitute humanistic expressions for objectivist ones would not be a real solution. Such a transformation would, first of all, undermine the identity—and ultimately the power—of the technical expert. Why would any writer willingly submit to being stripped of the hard-won resources of status and success? Humanists might argue that this kind of sacrifice is necessary for the project of communicative action in a democracy, that the technical experts may keep their ethos intact for writing that is directed to other specialists in their field, but for public documents like EISs, they should write to the average citizen in ordinary language; and if they find this impossible because of their habits of discourse, the BLM editors should do the job for them. Unfortunately, the complexity of rhetorical situations involving multiple discourse communities stands in the way of this simple solution. A century-old history of administrative action is not undone in a day (as President Carter came to realize when he tried to reform bureaucratic style in a sweeping executive order condemning gobbledygook—see Carolyn R. Miller, "Environmental Impact Statements").

In fact, democratic and capitalist values have shaped both the need for experts and the styles and structures of bureaucratic prose. To get a better sense of the overall structural limits of the EIS and their relation to the microculture of the BLM and the American culture at large, consider a single example of the documents we studied. The *East Socorro Grazing Environmental Statement* treats a topic of major environmental, economic, and social interest in the region—the leasing of public lands for private cattle grazing. It was on this very issue that in 1974 the BLM was first challenged in court under the National Environmental Policy Act and was forced to begin producing EISs (Bardach and Pugliaresi).

In the *East Socorro Grazing Environmental Statement*, the BLM proposes a plan that would significantly tighten the management of leased grazing areas to improve, in the long run, "vegetative production, density, cover, and wildlife habitat" and through the implementation of a system of resting and deferring use of pastures, to "eliminate competition of livestock with wildlife and wild horses for available forage" (U.S. BLM, *East Socorro Grazing* i). In addition to the "Proposed Action," the study considers five alternatives, which are given the following titles: "No Action," "Livestock Adjustment," "Pasture Capacity Level," "Enhancement of Sensitive Resource Values," and "No Grazing."

Effects of a Pseudo-Democratic Rhetoric

We quote at length the following passage to reveal how structural traits extend over several paragraphs. The passage shows that the EIS treats the successive alternatives according to the impact each will have on a given item of environmental concern. Typical of EISs we examined, this document, with its litany of repetitive structures, makes reading difficult for uninitiated readers. To render it all the more imposing, the passage is divided roughly in half by two high-density, page-length tables that offer numerical data in support of the claims made in the prose text. (The text itself might best be described as a table without the luxury of *x* and *y* axes.) Note that, instead of talking about people—the "human resource"—the EIS deals with "jobs" and "operators," that is, people reduced to their productive functions:

> In summary for the 20-year, short-term use period there would be trade offs between improving the quality and/or quantity of natural resources (e.g., vegetation, wildlife, wild horses, erosion condition, etc.) and decreasing the total number of livestock (Table 3-1) and total income (Table 3-1) and causing a decline in the quality of the ranching lifestyle of the operators in the ES [Environmental Statement] Area.
>
> After implementation of the Proposed Action a total of ten jobs would be lost. By the end of the short-term use period (20 years) all jobs could be regained and one new job would be created. Six subsistence small operators would be reduced to 0 AUMs [Animal Unit Months] after implementation of the Proposed Action. By the end of the short-term use period (20 years) two of these operators may reenter the livestock industry when vegetative conditions improve. The other four operators would have their grazing use permanently eliminated.
>
> After implementation of the No Action alternative a total of two jobs would be lost. By the end of the short-term use period (20 years) all but one job lost could be regained. The number of operators would not change from the existing situation.
>
> After implementation of the Livestock Adjustment Alternative a total of ten jobs would be lost. By the end of the short-term use period (20 years) all but one job lost could be regained. Six subsistence small operators would be reduced to 0 AUMs after implementation of the Livestock Adjustment Alternative. By the end of the short-term use period (20 years) two of these operators may reenter the livestock industry when vegetative conditions improve. The other four operators would have their grazing use permanently eliminated.

> After implementation of the Pasture Capacity Level Alternative a total of fifteen jobs would be lost. By the end of the short-term use period (20 years) all jobs lost could be regained and seven new jobs would be created. Six subsistence small farmers would be reduced to 0 AUMs after implementation of the Pasture Capacity Level Alternative. By the end of the short-term use period (20 years) two of these operators may reenter the livestock industry when vegetative conditions improve. The other four operators would have their grazing use permanently eliminated.
>
> After implementation of the Enhancement of Sensitive Resource Values Alternative a total of eighteen jobs would be lost. By the end of the short-term period (20 years) all jobs would be regained and nine new jobs would be created. Eleven operators would be reduced to 0 AUMs after implementation of the Enhancement of Sensitive Resource Values Alternative. By the end of the short-term use period (20 years) seven of these operators may reenter the livestock industry when vegetative conditions improve. The other four operators would have their grazing permanently eliminated.
>
> After implementation of the No Grazing Alternative a total of 22 jobs would be lost and not reinstated in the long term. Fifteen operators would be permanently eliminated. (U.S. BLM, *East Socorro Grazing* 3-1–3-4)

Two structural traits are responsible for the passage's lack of readability and the moral offensiveness experienced by many readers of the EIS. These are tied to the principles of objectivity as understood by government report writers: First, *objectivity implies impartiality*, which in American political life has come to mean "equal time." Structurally, this translates to parallelism with a vengeance. The redundancy of the passage is fully justified in this reductively rational approach to prose style. Second, *objectivity implies reification*. All people must be made into things that are countable in an operationalist logic. All effects upon nature and human beings alike and all relationships of person to person and person to nature must be stated in terms of increase and decrease.

The reifying function of objectification allows all potential impacts to be leveled and plugged into formulas. Considered in terms of increase and decrease, all effects of an action can be quantified and dealt with "equally." A decline in blue gramma grass can be entered into a formula, and a "decline in the quality of the ranching lifestyle" can be entered into a formula. This practice does not necessarily imply that the BLM is full of heartless bureaucrats who are as interested in the fate of blue gramma grass as they are in the fate of the ranchers in their district. The purpose of many of the management practices instituted by the BLM is to recognize the *interrelation* of blue gramma grass and cattle ranching, to protect the one by protecting the other. Reification permits a disassembly of the great chain of being in such a way that the privileged position of human beings is lost. It also permits the chain to be reconstituted as a *net*, the network of ecological relations recognized by modern science. But surely this discourse is not what the deep ecologists have in mind.

Reification[5]—which may be construed as a positive gain by the management expert or modern scientist—increases the negative rhetorical effect of the EIS for the readership that includes the small-time rancher dispossessed in the interest of environmental protection and future ranching. The EIS predicts that in the "short term"—in terms of natural ecology, twenty years is a very short time—*jobs* lost will be restored, but of course any reader knows that many of those *people* who hold the jobs at the time the EIS is written will be dead by the time their jobs are restored.

As a partial compensation for this rhetorical reduction, and in compliance with NEPA, the reified human subjects dealt with in this report were invited, along with other members of the general public, to respond to the draft EIS. The letters that appeared in response to the *East Socorro Grazing Environmental Statement* when it was released in draft form in 1979 restored to the communicative context much of the rhetorical heat lost in the reifying prose of the BLM resource specialists.

Voices from the Audience

In contrast to the objectivist prose of the EIS, the language of the letters is engaged and subjectively varied. The "small operators" who stand to lose the most in the proposed plan mainly argue that the BLM is undermining the free enterprise of the ranching business and ignoring the longevity of the those who would be hurt by the plan as well as the improvements they have made to the land on which they operate their ranches.

Complaining, in so many words, that the system has intruded upon the small rancher's lifeworld, one respondent writes: "I have reviewed the East Socorro Grazing Environmental Statement and found it very inadequate which could be described as a serious blow to ranching and private enterprise in these United States. Government bureaucracies have found another way to regulate our lives." (U.S. BLM, *East Socorro Grazing* 4-94) He continues with a rhetorical flourish, but inadvertently, probably unconsciously, picks up in a grammatical slip the reifying trend of the EIS, referring to himself by means of a relative pronoun usually reserved for things rather than people—"which" instead of "who":

> I am one of the smaller operators which is referred to on page 2-91, column 1, paragraph, H 4. It states that we are less dependent on income from ranching and therefore less effected by a reduction in grazing. In my case this could not be further from the truth, for I am totally dependent on my ranch for my livelihood. I can only see that from implementing the proposed action we smaller operators will be forced from our operation to be bought up by larger operations. This can only serve to continue to monopolize the cattle industry which at this time is badly in need of competition. No serious consideration of my situation was taken into account in the writing of the Impact Statement. (U.S. BLM, *East Socorro Grazing* 4-94; original spelling and grammar preserved as it is in the EIS)

A similar note, though less stoic and with more force of sentiment, is struck in an old woman's letter, which was dictated to her daughter who typed and mailed it to the BLM. Throughout the letter, she resists reification and even, in the third sentence of paragraph 1, humanizes or personifies a reified entity, addressing "BLM" as if the agency were an individual human being:

> (1) I am 81 years old and dependent on income from ranching for my livelyhood. With the proposed cut, I would not have anything. BLM what do you propose for me to do?
>
> (2) The mental anguish that this has imposed on me and my family is almost unbearable.
>
> (3) Throughout the 23 years we have lived on the ranch with Allotment No. 059, we have built fences and done thousands of dollars of

> S.C.S. work for soil erosion, dams, tanks for watering and pipeline construction. . . . all of this has been ignored by the BLM and with the proposed cut, we will have less than we started with 23 years ago.
>
> (4) We had a Management Plan and abided by it to the tee. We did nothing but improve our allotment through the years, spending thousands of dollars, time and energy to save the soil and the vegetation thereon. We are ending up in almost bankrupt conditions. This proposal has devaluated our ranch by $122,000.00. (U.S. BLM, *East Socorro Grazing* 4-95)

In paragraph 4, the voice changes rather dramatically, which indicates either more help from the daughter in composing this section or a cunningness in the rhetoric of the first three paragraphs (or perhaps a little of both). The sentimentality of the rhetoric at the start of the letter could well have been used to set up a quasi-sophisticated view of "free enterprise."

The shift to an economic argument aligns the letter with one from the Socorro Chamber of Commerce, which takes a less impassioned but equally resistant tone and represents not only the small operator but the cattle industry in general: "It is our understanding that the present carrying capacity for this area is 14,224 head of cattle and that the proposed carrying capacity will be 9,935 head; a cut of 4,289 head of cattle. This drastic cut will reduce the gross annual sales by $1,640,250. The overall effect of such a loss to the community will be staggering" (U.S. BLM, *East Socorro Grazing* 4-95). This letter also questions the methodology of the EIS, taking up the question of legitimacy and focusing on a problem that (as we shall see later) has also bothered critics of the EIS production process from the scientific community—the collection of data and the time frame of the report: "The decisions reached in the EIS were based on information gathered from a brief visit to each ranch. We feel the information should have been compiled over several years to show a trend in the industry" (U.S. BLM, *East Socorro Grazing* 4-95).

The perspective of the large operator is represented in a letter from one of the most successful large ranchers in the area who is also a leading banker in the town of Socorro. His interest is identical with the economic position defended in the Chamber of Commerce letter, but his critique, which emphasizes the inadequacy of the BLM's methodology, is far less forthright. He tries to take on the experts on their own turf and in their own terms. Though the author of this letter and others like him have dominated the economy of the region for many years, his rhetoric is that of one who is in the process of being dispossessed by the government expert as the most powerful figure on the range of central New Mexico. On the one hand, he complains that the language of the EIS is inaccessible to "the average layman"; on the other hand, his letter awkwardly mimics the stylistic features of that language in an unwitting (?) parody of the nominalized and passive style of the EIS,[6] suggesting an effort to distance himself from the "average layman" by showing himself fully capable of approaching the BLM not from a level of lower education but on equal ground while maintaining noblesse oblige in defending the average cattleman (who is his colleague and customer):

> In my opinion, the time necessary to prepare the EIS with the aide of maybe 100 technicians employed by the BLM to prepare the said impact statement of some 300 pages over a period of some two to three years in which the EIS is prepared in a technical manner and with wordage not easily understandable by the average layman is not enough. It is further my opinion that the time for putting input in concerning this statement is certainly not near enough—some two months to evaluate and be able to put constructive input into the hands of the BLM in the time allotted (some 50 days) falls far short of the time really required. I would say that at least a year would be necessary for enough study to be made by users of the public domain to put in constructive reasonable recommendations. A great many people feel that a monitoring of the actual operations of livestock on these public lands should be done for a period of years prior to making a determination of the carrying capacity. (U.S. BLM, *East Socorro Grazing* 4-100)

The perspective of business leaders like this one is easily dismissed—too easily dismissed, as the scientist Sally Fairfax suggests—as "self-interested or as motivated by a desire to continue their activities without consideration of environmental costs" (743). Thus the claims that inadequate time had been allowed for gathering of data (a scientific objection) and for adequate review and comment (a political objection) could be dismissed as based on the desire to continue present practice for as long as possible. But both the scientific and the political objections remain at least potentially valid.

A response from an officer of the Salado Resource Conversation District echoes the scientific objections with a cutting remark on predictability: "The constant predictions of vegetative response in 20 years is . . . questionable. When weather patterns can be accurately predicted over a period of time it might be possible to predict the exact increase in density and cover the E. [I.] S. keeps referring to" (U.S. BLM, *East Socorro Grazing* 4-113). The same letter raises the political objections and recommends an advisory board composed of "local ranchers, range specialists from New Mexico State University, Game & Fish Dept. representative, Soil & Water Conservation Bd. member, etc." (U.S. BLM, *East Socorro Grazing* 4-114).

Another letter cannily pits the conclusions of one government bureau against those of another. This comes from a representative of the tribe of Laguna Indians whose rangeland touches on BLM-controlled lands. "I find it difficult to understand a proposed reduction of 50% in cattle units on the data that the B.L.M. Technicians have acquired," he writes, adopting once again in the typical fashion of the dispossessed, the alien jargon of the dominant entity he resists. Over and against the 1976 finding of the BLM range survey that "the proper grazing capacity is 80 cow units," the letter cites a study conducted in the 1940s, in which "The Bureau of Indian Affairs Technicians . . . determined a 220 cow unit rating . . ." (U.S. BLM, *East Socorro Grazing* 4-106).

A handwritten letter from another Indian represents an entirely different perspective in an entirely different language. This is the perspective characterized in the EIS as that of the subsistence operator. He speaks not as a "concerned citizen" of the state, nor as a capitalist interested in maximum production, nor even as a member of a tribe, but as one who identifies his interest with that of the land:

> I am 72 years old and most of these years are spent in this area. I can truthfully say, there is never any cause to be alarmed about the range conditions out there. I have seen it at its best and I have seen it at its worst. But there never was a time I could be afraid of seeing it dry up completely. It's true our rainfall is not as heavy as other places in the state. Neither is it as frequent as other places. But when

> it does start raining we get plenty of it. Our range becomes beautiful. Forage for the cattle becomes plentiful. For some reason we are never given credit for what we have. They seem to forget a cow does not live on grass alone. They need variety like humans. (U.S. BLM, *East Socorro Grazing* 4-104–5)

The writer has great confidence in the endurance of the cow: "Gentlemen never underestimate an old cow, they can find grass and water anytime" (U.S. BLM, *East Socorro Grazing* 4-104). His confidence in the capabilities of the expert is not as strong; he trusts experience rather than learning, and yet he realizes his own status as radically dispossessed: "It is really the man on horseback, who ride the range day in and day out who honestly know the range. He doesn't need a book to get what he already knows. Neither does he need a college degree to learn the ways of an ole cow. . . . This is the man to talk to, He is the person who knows the land, the way he knows his hand—Listen to him. Maybe he can't even write, but he knows more than lot of people who claim to know it all . . ." (U.S. BLM, *East Socorro Grazing* 4-105).

Contemporary technologists and environmentalists alike—however much they may admire the music of this voice—would quickly mark the limits of their nostalgia at the assertion of optimism over the earth's ability to reproduce infinitely. To the modern mind, the "fate of the earth" (to recall the title of Jonathan Schell's provocative essay on the future of the nuclear age) is in the hands of the human race. The perspective of green politics and deep ecology, though heavily influenced by the Native American tradition, differs strongly in just this way: It insists on the ultimate responsibility of human beings to care for—indeed to save—the earth, a view that from the perspective of many native peoples would appear ludicrous if not irreverent.

Actions and Cultural Effects Resulting from the EIS

Part of the democratizing rhetoric of the EIS is that these responses to the draft are included verbatim in the final EIS. But the respondents' complaints about the lack of review time, the scientific deficiencies of the study, their own lack of technical expertise, and their disadvantage in confronting an army of BLM officials indicate their tacit recognition that their voices will have little effect—rhetorical or real.

They understand the contours of power as described by Jean-François Lyotard: "Access to data is, and will continue to be, the prerogative of experts of all stripes. The ruling class is and will continue to be the class of decision makers" (14).

The respondents' lack of access to and inability to influence the informational system on which governance depends is suggested by the condition in which their communications are reproduced. All of the errors in grammar and spelling in the letters are left intact; the letters are not retyped or printed in the same format as the rest of the document, but are left in whatever form of native roughness they were found in when the envelopes were opened at the BLM office, often handwritten or typed on typewriters with defective keys. The BLM includes no formal response to the individual letters. If anything, the inclusion of the letters in this form serves to glorify by contrast the technical quality of the BLM-produced portions of the document, like a rough drawing placed beside a classic portrait.

The "Proposed Action" to reallot the grazing permits, as set forth by the *East Socorro Grazing Environmental Statement*, was approved by an appointed Interior Department executive, and at the time of this writing, it is still in effect. In a 1988 telephone interview, the Environmental Coordinator of the Socorro Resource District of the BLM reckoned that as many as 95 percent of the estimated 150 proposed actions studied by his office each year are approved and put into practice. This estimate raises further questions about the connection between the EIS as a rhetorical form and the ostensible efforts to democratize the EIS procedure. Though in principle EISs "give environmental groups a legal and political instrument to cancel, delay, or modify development projects that they oppose" (Bardach and Pugliaresi 23), and though there have been some notable environmentalist victories based on court cases involving EISs (the defeat of the breeder reactor, for example, documented by Commoner in *The Poverty of Power*), in fact *ninety-five percent* of draft EISs—a percentage identical to the the estimated "success rate" given to us by the BLM official in 1988—escaped legal challenge in the first two years they were required, the very years in which they would most likely have been challenged (Bardach and Pugliaresi 23).

It is perhaps not surprising that fringe groups like small farmers and

Indians are marginalized further by government practices. But what is surprising is the degree to which both academic science and traditional capitalist interests (like the big-time rancher/banker we heard from in the *East Socorro Grazing Environmental Statement*) have come to feel increasingly dispossessed. Many mainstream scientists have claimed that no reliable scientific research is likely to be used in or precipitated by the production, reading, and discussion of EISs. The Canadian ecologist D. W. Schindler remarks acidly:

> Many politicians have been quick to grasp that the quickest way to silence critical "ecofreaks" is to allocate a small proportion of funds for any engineering project for ecological studies. Someone is inevitably available to receive these funds, conduct the studies regardless of how quickly results are demanded, write large, diffuse reports containing reams of uninterpreted and incomplete descriptive data, and in some cases, construct "predictive" models, irrespective of the quality of the data base. These reports have formed a "gray literature" so diffuse, so voluminous, and so limited in distribution that its conclusions and recommendations are never scrutinized by the scientific community at large. Often the author's only scientific credentials are an impressive title in a government agency, university, or consulting firm. The title, the mass of the report, the author's salary, and his dress and bearing often carry more weight with the commission or study board to whom the statement is presented than either his scientific competence or the validity of his scientific investigation. Indeed, many agencies have found it in their best interests to employ a "traveling circus" of "scientists" with credentials matching those requirements. As a result, impact statements seldom receive the hard scrutiny that follows the publication of scientific findings in a reputable journal. ("Impact Statement Boondoggle" 509)[7]

One group of scientific and social scientific researchers has argued that although "the use of scientific information is critical to integrated impact assessment" and although "the will to improve the use of scientific information in public policy-making does not seem to be lacking (for at least a substantial number of interested parties), no satisfactory method for achieving this aim has yet been developed. Results of even the most dedicated attempts to improve such use—most notably the environmental impact statement process—have been disappointing" (Hammond et al. 287). Other social scientists have criticized the methodology of EISs as "crude or blatantly inappropriate" (Friesema and Culhane 344). In short, mainstream science and social science do not influence the EIS, and the EIS is not likely to influence research as practiced within the scientific community.

Moreover, at least one legal specialist on the environment has argued that EISs "have little relationship to actual decision making. . . . Often they are done after basic development decisions have been made" (qtd. in Friesma and Culhane 339). In a probing and ultimately disturbing article, Bardach and Pugliaresi have argued that the intent of the law was not to offer information that would influence decisions, but that instead "the EIS was intended as an 'action-forcing' instrument, that is, a document that might demonstrate to all interested parties that the agency was in fact doing the mandated environmental analysis" (24). These authors conclude that the EIS has not succeeded in this objective, mainly because "agencies cannot be penetrating or creative when their analyses are directed and mobilized for primarily defensive purposes" (24). Rather than insuring the best use of land, then, the EIS becomes a means of proof, or certification, or demonstration. The EIS is not intended to inform action but to forestall action—legal action against the agency in question.

"Impact Assessment," as Rossini and Porter show, "began without theory or method" (24–25). In the short time since 1969, a full-scale "science" and "industry" have developed to fulfill the requirements of the law. Dreyfus and Ingram claim that the congressional authors of NEPA "never contemplated anything so extravagant as the multiple volume dissertations which are now commonly produced"; indeed, "NEPA made no provision for funding extensive additional work by the federal agencies" (256). The agencies have nevertheless absorbed the task and have grown with it as necessary. The NEPA manual on EISs that was distributed in draft form in 1987 (which uses the expanded title of Resource Management Plan/Environmental Impact Statement—RMP/EIS!) makes the process of "impact assessment" even more complex and demanding on the agencies. A strong case could be made that the agencies have responded to a perceived threat of legal action by producing enough paper to smother any efforts to contain or redirect bureaucratic action. In this sense, they have been quite effective—about ninety-five percent effective.

The typical approach of rhetoricians in the field of technical communication is to construe effectiveness according to the needs of a specific audience that reads for a specific purpose—and to assign responsibility for effectiveness to authors and editors. Cheney and Schleicher, for example, conclude in an article that treats the editing of EISs, "Because authors commonly fail to consider their audience and purpose, they turn out reports that their audience cannot use" (336). What is the purpose of the EIS? The overt purpose is certainly "to provide a lucid summary of a proposed project and its alternatives," and no doubt "that purpose is poorly served by the encyclopedism, obfuscation, and poor focus rampant in many EISs, which bury essential conclusions—or obscure their absence" (Luccitta, Schleicher, and Cheney 591). But what about the covert purpose or the purpose as seen from the vantage of the government bureau under siege by Congress and environmentalists? The main goal from this angle is to avoid legal action—a purpose that is perhaps poorly served by the humanist rhetorical tradition, which demands clearly stated conclusions and forthright earnestness in general.

Despite its success in achieving the purposes of a limited discourse community, the EIS holds little attraction for the project of imagining a discourse appropriate to communicative action within a broad-based environmentalist culture. It is, unlike many of the discourses we have examined, action-oriented, but it is rigidly adapted to narrowly defined instrumentalist actions and is devoted to control and exclusion. Over and against the wishes of Congress, it radically limits participation and diversity of interest, promoting the century-old cult of efficiency. In an attempt to create an "action-forcing document," the authors of NEPA gave an unheard-of priority to a written document. They tried to guide social action by creating a new genre of written discourse. But the cultural tradition of administrative bureaucracy has reasserted the privileged status of tradition over genre by coopting the EIS for its own covert purposes and in so doing may well have provided the conditions for the evolution of a new genre. As Walter Beale has suggested, "Not only do the characteristic motives and stylistic appointments of [genres] change over time; they are also subject to extension, transformation, and appropriation into alien territory . . ." (19).[8]

Rational World, Narrative World

A larger question remains. Can a truly democratic public discourse emerge among conflicting traditions and interests? The sociologist Richard Harvey Brown has argued the negative: "Political discourse presupposes some conception of . . . community in which knowledge for the sake of action can be communicated across groups, classes, and statuses. . . . such community and communication are today not possible. As revealed by a sociolinguistic and ethnographic understanding of class-limited speech codes, each group's world appears as an impenetrable mystery to members of other groups, even to people of good will who are seeking to help or to form alliances" (1). However, Brown offers a revealing suggestion about possible alternatives to this vision of fragmentation in his claim that "contemporary society is eviscerated of narrative form" (R. H. Brown 3). In a metaphor strikingly appropriate to our discussion, he argues that "the narrative social text has become an extinct or endangered species" (3). Certainly one of the difficulties in reading EISs is that there is no clearly revealed plot line, no unified, meaningful action represented. The stories of the ranchers, the land, and the animals are analyzed and reconstituted as high-density tables and line charts. Their futures are reduced to a series of blandly projected "options." But at least one rhetorical theorist—Walter R. Fisher, building on the work of Hayden White—holds out hope of an emergence or reemergence of a "narrative paradigm" to challenge the currently dominant "rational world paradigm" (3). It is worth noting that the seeds of narrativity survive within the EIS process—in the practice of inviting the people of a region to respond to the EISs that affect them. They often respond by reasserting their own stories against a discourse that through reification and analysis violates what might be called the narrative structure of their lifeworlds.

In chapter 6, we will extend this discussion of the relationship of narrative and social action by exploring some of the rhetorical correspondences between environmentalist novels—fictional narratives—and direct action groups like Earth First! and Greenpeace, groups that are living out stories suggestive of alternative paths in the environmentalist lifeworld.

[The] New World utopia, this promised land, was soon buried under the ashes and cinders that erupted over the Western World in the nineteenth century. . . . The rise of the centralized state, the expansion of the bureaucracy and the conscript army, the regimentation of the factory system, the depredations of speculative finance, the spread of imperialism, . . . and the continued encroachment of slavery—all these negative movements . . . brought back on a larger scale the Old World nightmares that the immigrants to America had risked their lives and forfeited their cultural treasures to escape. . . . [T]he mechanical New World displaced the "romantic" New World in men's minds: the latter became a mere escapist dream, not a serious alternative to the . . . new mechanical world which, with every fresh scientific discovery, every new invention, displaced both the natural world and the diverse symbols of human culture with an environment cut solely to the measure of the machine.

—Lewis Mumford,
Pentagon of Power (24)

What political utopians since the French Revolution have sensed is not that an enduring, substratal human nature has been suppressed or repressed by "unnatural" or "irrational" social institutions but rather that changing languages and other social practices may produce human beings of a sort that had never before existed.

—Richard Rorty,
Contingency, Irony, Solidarity (7)

6
Rhetoric and Action in Ecotopian Discourse

Extending Environmental Radicalism

During the 1970s, it began to be clear that, with the spread of environmental consciousness among the general public—perhaps indeed because of it—reform environmentalist groups like the Sierra Club came to seem less radical than they once had seemed. Part of the rhetorical strategy of James Watt and the Reagan administration in the early 1980s was to divert public sympathy away from the activist groups by cultivating the image of environmentalism as a "protest movement," a "special interest," a radical fringe of American life. As Samuel Hays has shown in his history of environmental politics, however, this rhetoric missed the mark; Reagan and Watt wildly underestimated the American public's growing acceptance of environmentalist values. As the public developed points of identity with activist groups, the groups themselves shifted to the center of the American political spectrum in an effort to form stronger political links and to develop an environmentalist hegemony, the chances for which, to this day, have increased with every new environmental disaster, thus raising the possibility of a culture infused with such environmentalist principles as limited growth, ecological holism, sustainability in technological planning and practice, stewardship of resources (including materials recycling and protection of air, soil, and water), biodiversity, and in some form, wilderness protection.

Not all environmentalists, however, were pleased with the direction the movement had taken. Many thought it had been compromised and rendered ineffectual by having been absorbed into the mainstream of American life. In particular, the fear that the general public's adoption of environmentalist values applied only to protection of personal property and individual interests—very likely a correct assessment—radicalized a number of writers and other figures connected with reform environmentalism. Wilderness protectionists were especially prone to become radical because their goals would not have been advanced very strongly by the middle-class public's appropriation of environmentalism and the corresponding shift of focus in the environmentalist movement toward urban interests in clean air and water and other necessities and amenities threatened by environmental degradation. To keep alive the spirit of the outsider in the environmentalist movement, the radicals began to press the movement toward more sweeping changes and stronger political commitments.

Emerging from reform environmentalism to challenge the public toward new heights of awareness, many of the utopian radicals of the mid-1970s banded together in experimental political action groups. The first such spin-off group, the Greenpeace Foundation, came to the public's attention early in the decade with its "save the whales" campaign, which employed methods of protest derived from Quaker social action philosophy and from the Civil Rights and peace movements. This approach to social change emphasized "bearing witness" against the wrongs of social practice, arguing and standing for the elimination of social acts deemed immoral or unjust, a clearing away of the weeds and deadwood to make way for new forms of social and political growth. Others followed Greenpeace in preserving the status of environmentalism as a protest movement, an effort to "negate the negative" in American social history. Nowhere was this outlook more evident than in Edward Abbey's provocative and mildly prophetic satire, *The Monkey Wrench Gang*, published in 1975. Much to Abbey's surprise (and barely concealed delight), his story of four friends who confront the developmentalist establishment with an intensive program of sabotage inspired a group of disaffected wilderness protection advocates to form Earth First! in 1980, a loose confederacy of radicals devoted to undermining ecologically damaging practices through methods ranging from theatrical displays that model acts of sabotage to the acts themselves.

The work of the new utopians was not totally dominated by the negative strategy of resistance politics, however. Acting on the archi-techtonic, or constructive, impulse of utopianism, Ernest Callenbach brought out *Ecotopia* in 1975, the same year Abbey's novel appeared. The novel projects a vision of environmentalist history as it might have developed if a region of the United States had seceded from the Union to form a society based on ecological values. In contrast to the rhetoric of other radicals, Callenbach's book makes a smoother, more subtle appeal to the general public, offering its eccentric views in the popular genre of science fiction where such views, if not expected, are at least tolerated by readers whose censoring mechanisms are relaxed in the presence of this "escapist" genre. Like the work of the other utopians, *Ecotopia* nevertheless urges an extension of environ-mentalist consciousness beyond a recognition of problems in the backyard and the favorite vacation spot. It is about rethinking our total and global relationship with the social and the natural world. With varying degrees of rhetorical power, Callenbach and the rest of the utopians make a significant contribution to environmental politics by creating extreme visions against which to measure the more modest goals of *Realpolitik* as practiced by reform environmentalism and the yet more modest goals followed by citizen recycling groups and Earth Day committees.

Greenpeace and the Rhetoric of Resistance

To project a sense of urgency and to stimulate the growth of a culture founded upon the tenets of ecological wisdom, the Greenpeace Foundation has taken a dramatic approach to a rhetoric of resistance—dramatic both in the sense that it has its origins in literary drama and in the sense that it has a strong effect on those who witness Greenpeace's staged "happenings," the mass media and an interna-tional public. With its sensationalist rhetoric of action, Greenpeace has successfully manipulated the media into covering stories that would be ignored if the ordinary principles of information value were followed. Where there are no events deemed reportable because they lack human interest and are not, in the strictest sense, "news," Greenpeace creates events; its members are "newsmakers" in the truest sense. Whale hunting in the seas of the far Northern Hemisphere, for example, holds little interest for ordinary media users in their comfortable suburban homes. But if in a man-bites-dog tactic, a group of romantic young heroes sails forth to interfere with the "evil" practice at great risk to their own lives, the story takes on a powerful new slant. Greenpeace campaigners have risked their own lives as a form of resistance to actions against the body of the earth and have thereby modeled a naturalized view of the relation of human beings to the earth, a shift in consciousness suggesting that human lives are always ultimately at risk whenever people, acting from a perspective that regards the human condition as alienated from nature, violate the ethical principles of ecological wisdom.

Bearing Witness

Greenpeace was founded in Vancouver, Canada, in 1969, an eventful year in the history of environmentalism, the year in which the National Environmental Policy Act was passed by the U.S. Congress. The group began as the "Don't Make a Wave Committee" of a local chapter of the Sierra Club, formed to protest underground nuclear testing in the Aleutian Islands. Drawing upon the practices of the civil rights and antiwar movements of the 1960s—the sit-ins, the staged arrests, the happenings—the committee landed upon the idea of sending boats and bodies to be present at the scene whenever a nuclear test was planned. If the test was to be completed, human lives would have to be taken. The first boat was christened *Greenpeace I* (Perlman 59).

Thus an immediacy was injected into the message that the protesters wanted to get across: Nuclear testing is ultimately deadly to life on earth. The rhetoric implicit in these symbolic actions eventually evolved into a full-blown ethic and method of protest, drawing upon but adding an active element to the classic Quaker concept of "bearing witness": "In the Greenpeace ethic, confronting means not only 'bearing witness,' but also 'presenting these atrocities to the world by taking direct actions to subvert them' " (Purl 26). The protest, to be considered effective, must not only draw attention to an unethical action but also bring these actions to a halt at least for the moment, much as a union strike will close down a factory.

Early on, Greenpeace's partnership with the news media—at times an uneasy alliance—was established in a distinctly modern version of the neo-Quaker ethic, involving a simultaneous subversion and exposure of unconscionable actions. As a reporter for *Rolling Stone* magazine put it, "The 'subverting' is up to Greenpeace; 'presenting' is up to the media"; "the witnesses would themselves be witnessed" (Purl 26–27). Stories that, before Greenpeace, seemed remote from the concerns of television watchers and magazine readers—above all, the overharvesting of endangered species of whales by Soviet, Japanese, and Icelandic fishing interests—came to be widely covered because of Greenpeace's efforts and in the process, captured the imagination of the media-saturated American public. Candidates for Miss America began to express their concern for the fate of the whales; cartoonists created characters sporting T-shirts with slogans like "Nuke the Whales."

Greenpeace thus achieved a high international profile through consciously seeking the attention of the news media. In addition to providing numerous opportunities for high drama, the Greenpeace organization courted newspeople with ready-made scripts, high-quality photographs and film, news releases, and extensive background reports, all informed by the heroic myths of deep ecology and enlivened by the daredevil antics of personable young protesters who win converts to their cause through their playfulness as well as their daring and personal commitment. The often whimsical tone they have taken toward their work contrasts strongly with the other more sullen, puritanical, or threatening faces of protest that emerged in the news of the 1960s and 1970s—not only in the American antiwar movement but also in the arena of international terrorism.

A *Range of Media Response*

Greenpeace's effort to interfere with whale harvesting in the mid-1970s was greeted with a flurry of media response, not all of which, however, was presented without irony or resentment. Four reports from popular magazines demonstrate the range of responses Greenpeace was able to elicit from the media, from deeply sympathetic to ironic to forthrightly hostile:

1. A report in *Oceans* of July–August 1977, covering a whale-saving expedition of July, 1976, typifies the sympathetic response as well as the use of high drama in magazine journalism. It begins this way:

> A black Russian catch boat bore down on the fleeing family of sperm whales whose tiring muscles were no match against the relentless engines. The harpooner readied and lowered his cannon.
>
> To his dismay, he found himself sighting on two defiant faces staring at him from a Zodiac inflatable boat. The nonviolent Greenpeace protesters had found their target.
>
> The Russian captain tried to speed past them; when that failed, he attempted to run them down. The protestors kept themselves neatly between the catcher vessel and its prey. The whales escaped. (Perlman 58)

The reader's lack of familiarity with the scenes and practices of whaling in the North Pacific is compensated by the familiar mythos of the classic melodrama—the villainous Russians relentlessly pursuing (in a black boat!) a leviathan "family" of victims, until the skillful, though less powerful, utterly admirable (nonviolent) heroes successfully foil the evil action.

From this point on, the melodramatic tone lifts somewhat as the reporter seeks the balance of realism. The story flashes back to the details of Greenpeace's evolution as an organization and includes a brief account of a less successful encounter of 1975, in which a whale was killed right before the eyes of the would-be saviors. The author attests as well to notable political failures that Greenpeace has faced: "Despite the outpourings of international indignation and increasingly vocal calls for an end to seal hunting"—which followed upon the efforts of Greenpeace to interfere with the hunting and to flood the media with pictures of cuddly baby seal victims—"the 1977 harp seal quota was raised!" (Perlman 61). Nevertheless, nothing is lost in the sympathy cultivated by the author for the Greenpeace cause, who finds that underdogs make excellent heroes.

2. A report in *Smithsonian* of August 1976 tells the story of the first effort of Greenpeace to interpose bodies between Russian whalers and their targets. Like the story in *Oceans*, this report recounts the romantic drama of "the Greenpeace adventure": "a crew of courageous

idealists on a creaky boat . . . interpose themselves to foil the harpoons" (Herron 23). But this report is more measured and less melodramatic, perhaps because the reporter himself was present, unlike either the reporter in the *Oceans* story who had to rely on the secondhand account of the president of Greenpeace ("Bob Hunter . . . laughed when he recounted the story" [Perlman 58]), or the author of a report in *Rolling Stone* who quotes at length from the journal of a Greenpeace member to fill in the details (and the rhetorical slanting) of the actions (Purl 28–33).

The *Smithsonian* reporter is clearly moved by the plight of the whales themselves. From a naturalized, eco-ethical perspective, he says of a harpooned whale, "She was crying exactly like a baby" (Herron 28). Of his first view of the Russian whale-processing factory ship, he writes, "It was a scene from a clockwork inferno: decks running red with blood and entrails, squeaks and groans from the donkey engines, smooth shapes of whales disintegrating into efficient piles of gore" (26).

But it is the system, the machine, of whale destruction that draws his environmentalist indignation, not the persons of the whalers themselves. He makes every effort to portray the Russian whalers not as villains, but as mere workers, cogs in the machine. The workers stand mute when the Greenpeace activists address the ship over a loudspeaker: "We are an international group of environmentalists who oppose the killing of whales. We speak for the 53 members of the United Nations that voted in 1972 for a ten-year moratorium on all whaling. We ask you to stop killing whales at once, to leave this business, to convert your fleet to other uses. If you do not, we intend to do everything within our means to prevent you" (26). The Russians may have failed to respond for a number of reasons. Perhaps there was a problem with the translation; perhaps they heard a threat in the last sentence, since nothing preceding it suggested the nonviolent nature of the protest. But the *Smithsonian* reporter provides a more sympathetic reading: "The Soviet crew reacted like almost any group of factory workers confronted with an unexpected political message—that is, they hardly reacted at all" (Herron 26).

Despite the reporter's clear sympathy with the environmentalist cause, he stops short of overtly heroizing the Greenpeace activists, creating distance through his efforts to grant equal sympathy to the Soviet workers and through his reference in the article's title to the Greenpeace campaign as "not altogether quixotic." This suggestion is followed up in a balanced assessment of the ultimate effect of Greenpeace's actions, in which additional distance is attained through the use of expert opinion: "Militant confrontation with whalers, as epitomized by the Greenpeace adventure, does serve to highlight the continued—but legal—slaughter of great numbers of whales by the USSR and Japan. This admission by Dr. Robert White, Administrator of the National Oceanic and Atmospheric Administration, does not mean that there is American sanction for such tactics, but it reflects interest in conservation of whales" (Herron 30). In other reports, the distance reflected here is compounded further by reporters who do not share the *Smithsonian* author's deeply environmentalist orientation.

3. A report in *Time* of 9 July 1979 is one such story. The author is quite aware of the mythic potential of his tale. We are given, for example, the source in Native American myth for the name of the Greenpeace ship *Rainbow Warrior,* which would become famous when in 1985 it was bombed and sunk in mysterious circumstances involving an encounter with French nuclear testers in the South Pacific (see Dickson; Simpson). The vessel's name was derived from "an old North American Indian legend that when the earth's animals have been hunted almost to extinction, a rainbow warrior will come down from the sky to protect them" ("Whale of a War" 45). But in telling the story of Greenpeace's encounter with an Icelandic whaling expedition, the reporter exhibits a strong sense of irony, which manifests itself in a portrayal of the environmentalist campaign as the unwitting cause of needless suffering: "For 18 hours, the stalemate continued. Every time the whaler angled close enough to the fins for a shot, one of Greenpeace's four inflatables would dodge into its way. The contemporary Ahab [Icelandic whaler Thordur Eythorsson] was forced to hold his fire lest he hit the protesters. Finally, after two misses, the captain got off his shot when one fin surfaced directly in front of the catch boat. It was a painfully slow demise for the beast;

to minimize the danger to the protestors, Eythorsson had removed the explosive cap from the harpoon" (45).

The report uses the Greenpeace story as a dramatic entrance to an account of a meeting of the International Whaling Commission to consider a temporary moratorium on whale harvesting. The irony continues: "Even if none of the moratorium calls are approved, the whaling industry may be sunk by dwindling profits. . . . Of whalers or whales, which will die first?" ("Whale of a War" 47).

4. An essay in *Field and Stream* of April 1986 exchanges irony for sheer outrage at Greenpeace's tactics, a predictable response from a magazine that advocates conservation causes but also represents the hard-line "law and order" ethic of the American hunter, drawing its readership not only from groups like Ducks Unlimited but also from the National Rifle Association. The author charges that "Greenpeacers . . . frequently resort to unethical and/or illegal means in trying to accomplish their self-righteous ends" (Reiger 14). He openly compares their manipulation of the media with the work of international terrorists—an analogy that is covertly suggested in other portrayals of the Greenpeace activists (often called, to the chagrin of the Quaker-bred group, "commandos" [Harwood 75]). Cannily, he also points up the inconsistencies in the "save the whales" campaign. The one species of whale that, according to this article, is in perhaps the greatest danger of extinction, the bowhead, holds no interest for commercial operations and is hunted almost exclusively by the Inuit people of Alaska. Greenpeace has said nothing about the plight of this whale because "a wildlife protectionist taboo" prohibits the suggestion "that any native people be limited in any way in their 'God-given right' to kill wildlife" (14).

The *Field and Stream* reporter, like his counterparts in the other magazines, is fully aware of the mythic status of Greenpeace. But he offers quite an original mythological frame for the environmentalists' campaigns. After criticizing the television program *60 Minutes* for using Greenpeace's prepared films and scripts in a profile story—"the show's producers rolled over and played lapdog to the Greenpeace proposition that terrorism is justifiable so long as one's heart is pure"—the reporter offers this interpretation of the media coverage of Greenpeace:

> The media learned long ago that John Q. Public finds due process boring. It may be the right way to express a grievance and make social adjustments, but court cases generally lack the drama of physical confrontations, particularly when one of the contestants is a diminutive Jack taking on a giant. Greenpeace portrays itself as Jack, and in an infrequently considered way, this is true.
>
> After all, Greenpeace justifies its trespasses against social law and order the same way Jack justified his theft of the magic harp and the golden-egg-laying hen, and later the destruction of the giant himself. Jack and Greenpeace both insist that since it seems unlikely that the meek will ever inherit the earth, the meek should seize it now, regardless of other people's claims to it. (Reiger 14)

After calling attention to the multimillion-dollar budget of the Greenpeace Foundation, the *Field and Stream* article concludes that instead of supporting such organizations, "People should send their $10 contributions to state conservation agencies"—many of which are, of course, heavily influenced by the hunting lobby (14).

The depth of the conflict between the position of the *Field and Stream* author and that of Greenpeace is suggested in the reporter's allusion to "Jack the Giant Killer." The psychoanalyst Bruno Bettelheim, in a study of the use and meaning of fairy tales, points to this story as a classic instance of the male child's fantasy to destroy the father and take the prizes of adulthood—money and fertility, conflated in the image of the hen that lays golden eggs—for his own. By telling such stories to children, parents suggest to children "that they can eventually get the better of the giant—i.e., they can grow up to be like the giant and acquire the same powers"; moreover, "if we parents tell such fairy stories to our children, we can give them the most important reassurance of all: that we approve of their playing with the idea of getting the better of these giants" (Bettelheim 27–28). If we translate "child" in this account to "political outsider," we can give the reading an ideological slant. By converting the fairy tale to a moral tale—from a vehicle for the release and sublimation of fantasy to a vehicle for repression and control—the *Field and Stream* reporter emphatically takes the side of the established order against those who would themselves grow into a position of control. He takes the part of the stern father who, in insisting on his own authority above all others, denies the child the ultimate right to participate in adulthood or citizenship.

The Effects of Greenpeace Actions

Greenpeace activists have put their bodies on the line both in the "save the whales" campaign and in their nonviolent protest against French nuclear testing, which resulted first in the severe beating of Greenpeace leader David McTaggart in 1973 and then in the death of a Greenpeace photographer in the *Rainbow Warrior* incident in 1985 (M. Brown 36). Their willingness to take risks has established a precedent whereby reporters know that, when Greenpeace calls, a dramatic story is imminent, one that readers will likely be interested in regardless of their political orientation. Greenpeace actions are mightily dramatic and also increasingly effective, according to a 1987 report in *Oceans:*

> Greenpeace members have landed in Siberia and challenged a Russian whaling ship in dinghies. They have parachuted from an Ohio smokestack to protest acid rain, plugged a chemical-effluent pipe in New Jersey, and attempted to plug radioactive-waste pipes in England. The group was a major force in halting the dumping of radioactive materials in the oceans. It was also instrumental [finally] in reducing the annual slaughter of Canadian harp seal pups by 90 percent and in pressuring the International Whaling Commission to vote a moratorium on commercial whaling. (M. Brown 36–37)

Many would argue that the presence of Greenpeace protest in several of these actions is incidental to their outcome. The *Time* article on the International Whaling Commission as well as other articles in *Oceans*, for example, have noted that the economic basis for whaling has so eroded that the moratorium is largely meaningless. The whale ships encountered by Greenpeace were in disrepair; the industry was clearly dying as synthetic substitutes for whale products were creating market conditions that undermined the profitability of whaling ("Whale of a War"; Perlman).

But other reporters, and not just those writing for *Field and Stream*, have gone further in their criticism than merely suggesting that

Greenpeace is ultimately ineffective. They have begun seriously to question Greenpeace's motives and methods. Even colleagues in environmentalism have grown wary. As one of Greenpeace's information officers admits, "Depending on whom one asks in the U.S. conservation community, Greenpeace is either one of the most effective environmental groups or one of the most irresponsible" (Dykstra 5).

Likely, Greenpeace has been both irresponsible *and* effective on occasions. But from the perspective of rhetorical analysis, the pressing questions are, how effective has the group been, and can it remain so? And has the appearance of irresponsibility hurt its chances for influence? Certainly Greenpeace's confrontational tactics have done little to heal the wounds of the environmental debate; the campaigns are aimed at converting not the enemy (despite the pleas to the Soviet whalers to channel their work to other ends) but rather to the masses that have remained neutral on environmental issues. As a reporter in *Science* wrote of the *Rainbow Warrior* incident, "The affair has hardened attitudes on both sides to the continued testing of nuclear weapons by the French government" (Dickson 949). Greenpeace's efforts to manipulate the media in the wake of the incident, however, have been part of their overall aim to get the attention of people who have not taken sides. But has the novelty—and thereby the *newsworthiness*—of Greenpeace worn off; and is the power of the crusaders to manipulate the media, the very source of their effectiveness in winning converts to the environmentalist cause, wearing thin?

In a recent article in the *New York Times Magazine*, Michael Harwood, a journalist with an impressive record of covering science and environmental topics, reflectively probes the aims of Greenpeace's actions. In many ways, the article is typical of most magazine coverage of Greenpeace. It gives the history of the group, charting its course from its small beginnings in protest of nuclear testing, through diversification into issues like whaling and the seal harvest and most recently toxic wastes, culminating in a portrait of an unwieldy international organization, linked by a sophisticated computer system, outfitted with a "navy" of ships carrying equipment capable of sensitive scientific measurements, but troubled by squabbles among the young idealists who drive the Greenpeace campaigns. Right down to the portrait of internal dissent, which in many ways fits nicely into the cultivated image of Greenpeace as a loose collection of unyielding young idealists, a group of "strong-minded individuals" whose projects are "generated at the bottom of the organization" (Harwood 76), the story has the feel of Greenpeace's official version of its own history (compare M. Brown; Perlman; Purl; Dykstra). The impression is not dispelled by the use of photographs provided by Greenpeace. Clearly the author has fallen under the Greenpeace spell. He opens his story with an account of Greenpeace campaigners perched on a crumbling rock cliff in Quebec to hang a banner urging industries in the area to save the whales in the Canadian tidal rivers by ceasing to dump industrial wastes. Harwood witnesses the action from the deck of a Greenpeace ship and is clearly concerned about the safety of the young campaigners: "It was all very picturesque and dramatic, but everyone was edgy, because loose rock was tumbling off the cliff near the climbers, who were suspended several hundred feet up" (72). But Harwood works hard to clear his head and look critically upon the performance:

> When I looked around at the handful of people actually observing the action, my first inclination was to ask, "Who's it *for*?" So far as I could tell, no one in the town on the other shore could see what was happening, nor could the people on ferries a mile or so away. Even the Canadian Coast Guard paid little attention. The dangerous stunt seemed to be taking place more or less in a vacuum. What good did it do?
>
> But it didn't take place in a vacuum. When the climbers were fighting the windblown banner, photographs and video images were being made, story notes scribbled and live radio reports transmitted. Greenpeace is always prepared to provide its own photographs to picture editors and it has the facilities to distribute edited, scripted and narrated video news spots to television stations in 88 countries within hours. (Harwood 72)

Even as he participates in the activity, Harwood is thus aware that Greenpeace is using him as an intermediate audience, the ultimate audience being the readers and viewers of the news reports. Harwood takes a pragmatic approach in his criticism of this rhetorical activity, questioning its effectiveness rather than its ethics: "The bottom line, of course, is what has Greenpeace accomplished?" (76)

This is not easy to determine, Harwood rightly contends. There have certainly been major breakthroughs in the areas of Greenpeace protest—the French shifted to underground nuclear testing, the harvest of whales and seal pups has radically declined, and several European countries have banned imports of white seal fur (Harwood 75)—but other environmental groups have worked on the same problems, using the "gentlemanly, legal-channels approach" (74), so the effectiveness of Greenpeace cannot be measured by the apparent outcomes of the issues they select to work with. "It is obvious, nonetheless," Harwood concedes, "that Greenpeace's contributions to publicity, particularly the images of slaughters and confrontations between 'good guys' and 'bad guys,' have been an important factor" (76). He quotes David Rapaport, a Greenpeace manager: "We have a tremendous ability to inspire. That's probably the biggest thing we can do"; the actions "can get the political process moving" and "can get the regulatory process moving" (76). Pierre Béland, a Canadian biologist agrees: "You may have seen scientists for years saying something—with proof—and no action is taken, [government] ministers don't react. Then Greenpeace does something for a few months, and suddenly you have ministers asking their scientists, 'Why haven't you worked on this?' " (76).

Greenpeace gets attention; therein lies its success. Although the reception of the magazine press has varied and although Greenpeace methods have been frequently criticized or treated with irony, the actions have certainly not been ignored. The effect has been to bring international attention to selected issues of environmental concern. In this way, Greenpeace contributes to the overall environmentalist effort. The Greenpeace campaigns do not appear to be as goal-oriented as most political actions are. If their effects are diffuse and even off the mark, however, little is lost in the way of public attention. Even a negative report will get people thinking. So what if green dye on the fur of a baby seal makes it more vulnerable to natural predators so long as the act of dyeing the fur will ultimately lead to an alteration in the human conscience that will save the seals from their most efficient and devastating enemy, the human hunter?

Though Greenpeace leaders like to describe their work as involving "direct actions," these actions are at best indirect. Moreover, they are not *sustainable*, to use Lester Brown's terminology; they are flashes in the pan. They model commitment and concern, but they do not suggest a set of actions that ordinary people can take up as a daily practice to improve ecological and economic conditions. To plug the pipes of a polluting company no doubt says to the public, "You too should work to stop this action," but it gives no indication of how that is to be done or what practices might serve as alternative actions—what kinds of jobs the local people will have, for example, after the plant is closed. Rather than reading Greenpeace as a failure in modeling actions, as either a true political action group or a reliable source of action-oriented information like the Worldwatch Institute, it is more flattering, and probably more accurate, to read Greenpeace as a *consciousness-raising* and *conscience-forming* rhetorical force.

Greenpeace and Information

In recent years, however, Greenpeace has tried to add an information dispersal function to its primary rhetorical function as a group devoted to awareness-raising and conscience-molding. The organization now publishes a magazine and has increased the amount of research that accompanies actions. Information about their research is distributed in the form of reports to press outlets and local environmental groups as follow-ups to the actions. This move is likely the result of Greenpeace's expanded interests. While the tactic of interposing bodies is most effective in protesting whale harvesting and nuclear testing, it is not as effective in protesting the release of toxins into rivers, oceans, and the atmosphere. Remaining true to the strategy that has worked in the past, the activists have hung banners and have skydived from smoke stacks. Some have even been known to allow the poisonous effluents of chemical factories to pour over their bodies. But this practice does not have the immediate effect achieved by interfering with whaling or hunting and may seem just plain foolish since the effects of such poisons may not exhibit themselves for several decades, by which time the issue will probably have been settled one way or the other. Since the favored strategy of the group has proved to be limited, therefore, the activists have turned to more conventional methods of civil disobedience and rhetorical combat. So far, they have met with only very limited success. Although they have forced developmentalists in the pesticide industry to respond publicly to information packages released to the press and to other environmentalist groups, the ensuing exchange—typical of a world dominated by ecospeak—has degenerated into a propaganda war.

Consider, for example, the case of Greenpeace's campaign against the dumping of toxic wastes, once again a global issue but also a major problem in the interior United States that has brought the Greenpeace fleet to the inland rivers. Greenpeace communications officer Peter Dykstra describes the campaign thus: "An international campaign calling for 'source reduction'—legislative and technological schemes to reduce the amounts of hazardous waste generated by industries—serve as a foundation for Greenpeace's international work on toxic wastes. Methods range from conventional lobbying and community organization to literally plugging the pipes of major polluters" (Dykstra 45).

One target of this campaign has been Velsicol Chemical Corporation, which as we have seen, was treated as a major villain in *Silent Spring* and consequently tried to stop the publication of the book by accusing Carson of scientific irresponsibility and of participating unwittingly in a communist plot. Velsicol is still producing and exporting heptachlor and chlordane, two controversial pesticides that Velsicol, in consultation with the Environmental Protection Agency and under the threat of an impending ban, pulled off the U.S. market. In addition to trespassing on the premises of Velsicol's plant in Memphis, Tennessee, to hang banners in protest of this activity and thereby to create an impressive photograph opportunity for local reporters, Greenpeace also issued a follow-up report, "Exporting Banned Pesticides: Fueling the Circle of Poison," billed as "A Case Study of Velsicol Chemical Corporation's Export of Chlordane and Heptachlor." The ultimate aim of the report is to influence the legislative process, specifically to close "the loophole in federal law which permits 'for export only' production of pesticides which cannot be used in the United States" (Marquardt 1). To achieve this effect, the report draws upon a wide range, indeed an eclectic set of sources, including a good number of papers reporting the results of solid applied scientific research on the toxicity of the pesticides in question.

In addition, an appendix to the report gives a set of uninterpreted and unrefereed measurements taken by Greenpeace's own staff members aboard the *Beluga*, the organization's floating laboratory, during its campaign on the Mississippi River. With the review of the literature and the presentation of raw data from original research, there is a clear claim to scientific legitimacy made in the report.

Against this claim, Velsicol directed its response in a five-page press release issued the same week as the report. The release asserts, "Greenpeace is trying to lead the public to believe that they have compiled a scientific report. In reality, their report has absolutely no basis in science"; the report is "fueled by emotion," contains "footnotes that do not substantiate their claims," and presents expert opinions that "in court proceedings have been labeled 'legally incompetent,' 'fatally flawed,' and 'insufficiently grounded in any reliable evidence' " (Velsicol). Velsicol thus demonstrates that the Greenpeace report depends upon, and is itself, "gray literature" rather than refereed scientific literature; draws unwarranted conclusions from applied scientific research; and in general, refuses to respect the limits of scientific data, applying findings freely to human as well as natural conditions. In short, Velsicol shows, quite correctly, that the Greenpeace report—whatever we may say about its rhetorical power to influence legislators and voters—is not a scientific document. By claiming, even implicitly, to have scientific status, the report opens itself to such criticism.

As a foil, the Velsicol press release contains a "Fact Sheet" that the company claims "is substantiated by peer-reviewed scientific information and U.S. EPA documents in the public realm." The Fact Sheet then trots out its own "scientific" evidence in a point-by-point refutation of the Greenpeace report. Nothing that it reports has any more scientific validity than the claims of the Greenpeace report. What it presents as "facts" are rather *findings* or *conclusions* of scientific research, even if the papers cited have been reviewed and published in the proper channels of normal science. The one fact in this entire case is that there are as yet no scientific facts on the toxins with which we are dealing. There are contradictory findings and conclusions. The issue is still warm and unsettled in the scientific community. Whichever side you happen to be on in the dispute, you will be able to find scientifically developed information to "support" your claims.

By drawing Velsicol into this press war, Greenpeace has, of course, brought public attention to the company's practices, and it has, once again, had its effect on the community consciousness, receiving media coverage in the local papers and on the local television stations. But has it hurt the environmentalist cause when it provides the opportunity, not just for Velsicol, but also for one of the bureau chiefs of the Food and Drug Administration to refer to the Greenpeace claims as "hype" (Brosnan A13)? It has certainly added credence to the frequently cited criticism of the environmentalist ethos—the charge that environmentalist appeals to science are unscientific and irresponsible, that they are highly selective and based on gray literature and other uncertified data.

Greenpeace's tactics represent a rhetoric of engagement and confrontation, which like the agonistic rhetoric of Rachel Carson and the editorialists in the Sierra Club *Bulletin*, represent an early stage in the political process. With the coming of age of the environmental movement, that stage may be passing, and the challenge for Greenpeace and the environmentalists becomes a matter of discovering how to move to the next stage or indeed what the next stage will be. In its efforts to extend research and to offer in its reports alternatives to environmentally damaging actions—"excellent results have been obtained by the use of cultivational techniques such as trap and barrier crops, crop rotation, field sanitation and the use of suitable plant varieties" (Marquardt 24)—Greenpeace campaigners are struggling toward a realization of a new goal. In this move, they overlap with organizations that have already matured in this discourse practice, such as the Worldwatch Institute. The early history of Greenpeace may well have hurt its chances to contribute to the mature stage of environmental protection. For who will take seriously a scientific report issued by a group whose very name is associated with life-risking commitment and change at any cost?

Deep Ecology, Earth First!, and the Rhetoric of Monkeywrenching

In its dramatic "actions," especially in the "save the whales" campaign, Greenpeace models a way of life that exchanges anthropocentrism, or human-centeredness, for biocentrism, a view of life that subordinates human needs to the needs of planetary life as a whole. Implicit in Greenpeace's perspective is the ethic of "deep ecology," the principles of which were articulated clearly in 1984 by the Norwegian philosopher Arne Naess (who coined the term "deep ecology") and George Sessions, an American philosopher and environmental activist. The main premises Naess and Sessions formulated are these:

> 1. The well-being and flourishing of human and nonhuman Life on Earth have value in themselves (synonyms: intrinsic value, inherent value). These values are independent of the usefulness of the nonhuman world for human purposes.
>
> 2. Richness and diversity of life forms contribute to the realization of these values and are also values in themselves.
>
> 3. Humans have no right to reduce this richness and diversity except to satisfy *vital* needs.
>
> 4. The flourishing of human life and cultures is compatible with a substantial decrease of the human population. The flourishing of nonhuman life requires such a decrease.
>
> 5. Present human interference with the nonhuman world is excessive, and the situation is rapidly worsening.
>
> 6. Policies must therefore be changed. These policies must affect basic economic, technological, and ideological structures. The resulting state of affairs will be deeply different from the present.
>
> 7. The ideological change is mainly that of appreciating *life quality* (dwelling in situations of inherent value) rather than adhering to an increasingly higher standard of living. There will be a profound awareness of the difference between big and great.
>
> 8. Those who subscribe to the foregoing points have an obligation directly or indirectly to try to implement the necessary changes. (Devall and Sessions 70)

On the map of perspectives developed in our introduction (figure 1, page 11), deep ecology stands in opposition to the views of nature as object and nature as resource.[1] Though deep ecology—above all an *ethical* movement and a variant of naturalism—is not necessarily connected with the idea of the sacredness of nature (Manes 145–46), we use the phrase "nature-as-spirit" to signify the pole of our continuum that deep ecology shares with nature mysticism, the impulse to merge the human self with the land in an all-consuming oneness, more or less the equivalent of Naess' "Ecological Self" (Manes 148), a concept with philosophical roots in Emerson's notion of the oversoul.

In dramatizing the radical biocentrism of deep ecology, Greenpeace has nevertheless kept open the possibility for forming links with more moderate conservation groups as well as groups with distinctively different viewpoints, such as organized labor.[2] From the perspective taken by the most extreme environmental radicals, Greenpeace has bartered away its radical status in this search for a broad-based environmentalist hegemony and has thus caused the most radical of its members to form spin-off groups. In 1979, the Sea Shepherd Conservation Society emerged from Greenpeace to carry out an angry campaign that has included ramming and sinking whaling ships in the Atlantic; and in 1982, the founders of Robin Wood left Greenpeace out of disappointment over the pace and tone of its radicalism and have since occupied smoke stacks to protest acid rain and have simulated land slides to demonstrate the effects of deforestation in Germany (Manes 17–18, 57, 125).

Earth First!, which has become the best-known model for radical environmentalist groups in America, came forth under similar circumstances. Frustrated with the inability of the Wilderness Society to accomplish its goals in Washington and angered by the deeply entrenched earth-as-resource perspective of government bureaus and conservation groups alike, the Earth First! founders adopted a no-compromise deep ecologist position based on a refusal to be appropriated by mainstream political interests and human-centered environmentalist projects. Where nonviolent protest and theatrical diversions fail to accomplish a local purpose, Earth First! encourages—though rarely takes responsibility for—acts of "ecotage" or "monkeywrenching," the purposeful sabotage of the machinery of developmentalism. To avoid what its members see as the creeping enticements of bureaucracy and conformity, the group resists adopting the trappings of an organization and remains a "disorganization" with no official leaders, no bylaws, no policy statements, no dues, and no centralized control of any kind (Kane 100). Out of member contributions, it maintains a newsletter, prints how-to manuals on resistance techniques, organizes demonstrations, and pays a subsistence salary to a few members who carry the message of radical environmentalism around the country.

Earth First! shares with recent theorists of deep ecology a radical desire and a psycho-semantic need to distinguish their perspective from that of other environmentalists. Both the Earth First! radicals and the deep ecologists abandon the appeal of reform environmentalism to the established mores of American life and try to cultivate instead a new language and a new action agenda for an ever-deepening understanding of the human relationship to the natural world. Dave Foreman, one of the group's founders and most articulate leaders, says that the mission of Earth First! is to provide "a productive fringe" where "ideas, creativity, and energy spring up" and "later spread into the middle" ("Earth First!" 41). In addition to producing energy, however, Foreman and his colleagues have discovered that the search for originality can consume energy as well, demanding constant vigilance against compromise and appropriation and necessitating potentially endless dislocations leftward. The history of Earth First! has thus been characterized by a frenzied commitment to action, usually action, which though far from thoughtless, lacks a clear connection to an established body of ethical or political principles—but action at all costs. "Action is the key," Foreman has said, "something more than commenting on dreary environmental impact statements" ("Earth First!" 42). The group's engagement in uncompromising actions has alternated with an equally dramatic public search for an ideology or political identity that matches the radical need to break through the stalemates of policy discussion and diplomacy.

Earth First! The Action and the Image

The search for an effective public image has preoccupied the leading voices of Earth First! from Dave Foreman to Christopher Manes, the young editor of the Earth First! newsletter and the author of *Green Rage*, a recent book on environmental radicalism. The case of Foreman, who has been with the group from the start, is of particular interest to our study of how the radical impulse transforms rhetoric and action. In his earliest writings, Foreman was never satisfied with attributing the emergence of Earth First! to his own inability as a lead lobbyist with the Wilderness Society. However, in that position, he had failed to influence the bureaucrats of the Carter Administration, whom he and his fellow environmentalists had considered to be their proverbial friends in high places. Though certainly forthright in documenting these failures, he wanted to blame the losses and the continuing entrenchment of the government's own brand of conservative conservationism on the movement of reform environmentalism as a whole.

In an article published in *The Progressive* in 1981, a year after forming Earth First!, Foreman appealed to history in an effort to bolster this position. "The early conservation movement in the United States," he wrote, "was a child—and no bastard child—of the Establishment" ("Earth First!" 39). Early conservationists were all an "elite band," he claims, lumping in eccentrics like John Muir and innovative thinkers like Aldo Leopold with the likes of Theodore Roosevelt and failing to mention altogether the Sierra Club's infamous "archdruid" David Brower, who was later to become one of the leading supporters for Earth First! Foreman argues that "it was not until Earth Day in 1970 that the environmental movement received its first influx of real anti-establishment radicals as anti-war protestors found a new cause—the environment. Suddenly, in environmental meetings beards appeared alongside crew-cuts—and the rhetoric quickened" ("Earth First!" 39). Ignoring the work of Greenpeace, founded in 1969, a year before Earth Day, Foreman continues in this pseudohistorical vein, basing his interpretation ultimately upon his personal history, his own conversion to reform environmentalism and his reconversion to militant nonconformity: "The militancy [of the environmentalist groups] was short-lived. Along with dozens of other products of the 1960s who went to work for conservation groups in the early 1970s, I discovered that a suit and tie gained access to regional foresters and members of Congress. We learned to moderate our opinions along with our dress" ("Earth First!" 39).[3] Realizing, however, that "the anti-environmental side" was able to win its battles by using "extreme, radical, emotional" tactics against the "factual, rational" arguments of Foreman and his associates in the Wilderness Society, Foreman determined to fight fire with fire. He reasoned that he and his fellow lobbyists had become "indistinguishable from those we were ostensibly fighting" ("Earth First!" 40), that the environmentalists had wrongly assumed that they could fight the system from within the system. Ironically, the rhetorical strategy Foreman finally adopted moved him closer to the "extreme, radical, emotional" practices of his opponents. Likewise, in abandoning the genteel politics of the "raging moderates" in the Wilderness Society and Sierra Club, and in finally advocating the politics of sabotage, he had decided to turn the violence of the "machine" back upon itself ("Earth First!" 41).

Soon after issuing this early manifesto, Foreman met Bill Devall and discovered in deep ecology a way of thinking that came closest to fitting the viewpoint of Earth First! In a 1985 interview with *Mother Earth News*, Foreman was even citing the work of Muir and Leopold, whom he had all but dismissed as elitists in 1981; his conversion to deep ecology reconciled him to accepting the mystically inclined reformers as the forefathers of Earth First! ("Consistency is the hobgoblin of little minds," Foreman had told an interviewer from *Audubon* in 1982. "That's the only thing Emerson ever said that's worth remembering" [Kaufman 118].) In the interview with *Mother Earth News*, Foreman describes his understanding of deep ecology: "Deep Ecology . . . goes back to John Muir's central insight that the human animal is only one of several million species of living things sharing the limited resources of this planet. And further, even though we are more intelligent and more powerful, we have no preordained right to totally modify, develop, and otherwise exploit the entire planet for our own use and whim" (qtd. in Petersen 18). The main attraction of deep ecology for Foreman was that it provided him with an ideological means of distinguishing the perpetrators of wilderness protection, who defended the land against the encroachments of civilization, from many of the enemies of Earth First!—the ranchers of the mountain West, for example—who also claimed closeness to the land and the right to defend it against the intrusions of big industry and big government. According to the logic of deep ecology, these cattle owners had forfeited their claim to oneness with the earth and ownership of the land by allowing overgrazing and other ecologically damaging practices. Their concern with resourcefulness had limited their understanding of the workings and worth of nature.

The ideological connection of this viewpoint with ecological sabotage or "monkeywrenching" remains unclear in the interview in *Mother Earth News*, however. When asked how Earth First! has applied the principles of deep ecology in successful attempts to influence environmental politics, Foreman describes a campaign whose techniques could have been borrowed from Greenpeace, even the Sierra Club. In 1983 Earth First! had provided "nonviolent tactics training" for volunteers willing to block construction of a new logging road in Oregon's Siskiyou National Forest. The blockade "bought time and slowed construction on the road, so that the Oregon National Resources Council and Earth First! were able to file a lawsuit based on the Forest Service's Rare II Environmental Impact Statement." The suit was a "resounding" success, blocking continued construction of the road. But all too quickly, the victory threatened to turn sour when a year later, President Reagan signed a bill that reopened the possibility of road building in the area. By this time, however, the Forest Service was reluctant to act. Foreman interprets their hesitance this way: "What it amounts to is that our blockade was so well received by the public, especially in Oregon itself, that the Forest Service is shying away from another confrontation with us." Though he fails to acknowledge fully the group's need for coupling protest with legal action, Foreman has good reasons to claim that the campaign to save the old-growth forest in Oregon provides "conclusive evidence of the power of nonviolent direct action" (qtd. in Petersen 19–20). Other Earth First! actions—hanging banners on dams, suspending people in trees to prevent logging, plugging the sewer outlets of chemical companies—also recall the work of Greenpeace and seem to represent quite well the ethics of deep ecology that puts human life on a par with the rest of nature.

By the end of the 1980s, though, Foreman began to show signs of weariness over the theatrical practices of soft radicalism and the kind of people drawn into Earth First! after the incorporation of deep ecology into the group's public persona. He seems to have grown tired of being labeled "the clown prince of the environmental movement" and of providing "comic relief" for the serious efforts of other groups (Kane 101). As early as 1982, he told an interviewer that "if we can involve humor, we will. But as far as being the clown princes of [the] environmental movement, that's over" (Kaufman 120). His humor could turn dark, and his rhetoric could become inflammatory, as when in response to a call for moderation and reason in a discussion of wilderness protection at a meeting sponsored by the Colorado Open Space Council, he responded with this analogy: "You walk into your house, there's a gang of Hell's Angels raping your wife, your sister, your old mother. You don't sit down and talk balance with them, you go out and get your twelve-gauge shotgun and come back in and blow them to hell" (Kaufman 119). By 1987, the hard-line radicalism of Earth First! was symbolized by, if nothing else, the number of arrests that the members had logged and by the FBI's ever-increasing interest in Foreman's activities.

Just as galling as the comic image, which Foreman felt underestimated the radical verve of Earth First!, was the association of the group with the kind of New Age rituals and Native Americanism apparently condoned by Devall and Sessions in their book on deep ecology (though later dismissed by Sessions; see Manes 145–46). When an interviewer from *Smithsonian* asked Foreman in 1989 about the role of the Council of All Beings and Gaia-worshipers in Earth First!, he said, "That's the woo-woo stuff. It's beyond me. But the diversity's good" (qtd. in Parfit 196–98). Other old-time Earth First! radicals were more forthright about their disenchantment; Roger Featherstone, for example, told the *Smithsonian* interviewer, "Until '86 we were glad to be rednecks. Now everybody's a vegetarian, eating soy cheese" (qtd. in Parfit 198).

Though in *Green Rage* Christopher Manes still proclaims that Earth First! is founded upon deep ecological principles, and though deep ecologist Bill Devall continues to embrace Earth First! as the action arm of the movement, Foreman seems to be undergoing another reconversion to militant nonconformity. His sympathies attach less and less willingly to the "earth-bonding rituals" and the "woo-woo stuff" of many deep ecologists. His greatest personal loyalty has always been to the kind of people that formed the original core of Earth First!—the disenchanted "buckaroos" of the Wilderness Society, the "rednecks for wilderness," and the lunatic fringe of deep ecology. Included among them is Darryl ("Feral Darryl") Cherney, who has said, "Once you start thinking biocentrically, you start to see everything in a new light. This is not a napkin, it's a tree. The language becomes different. If you want to call someone something bad, call him a human" (qtd. in Parfit 198).

Foreman has occasionally appealed to other groups of social radicals. "Obviously," he wrote in 1981, "for a group more committed to Gila monsters and mountain lions than to people, there will not be a total alliance with the other social movements. But there are issues where Earth radicals can cooperate with feminist, Indian rights, antinuke, peace, civil rights, and civil liberties groups" ("Earth First!" 41). This interest in hegemonic links, however, along with his pride over the ability of Earth First! to influence the public, has always been subordinated to his nonconformist, antinomian passion for the individually motivated and furtive righteousness of the radical acting alone or in small gangs of fellow seekers.

Herein lies the appeal of monkeywrenching. The great inspiration for Foreman and his favored circle of self-proclaimed misfits was and continues to be Edward Abbey's *The Monkey Wrench Gang*, the 1975 novel that symbolized the mid-1970s disenchantment with the pace and direction of reform environmentalism (Foreman, *Ecodefense* 1). The only political commitment of Abbey's characters, their only social "theory," is to preserve the homeland against the "machinery" of government, business, and technology. Abbey's characters—and Abbey himself—would agree with Foreman in this view, as the novelist made clear in what was likely the last interview he gave before he died in 1989: "When someone invades your home," he said, "you don't respond objectively and reasonably. You strike back with emotion, with rage. Well, government and corporations are invading the wilderness, our native natural home. There's no time to be dispassionate about that" (qtd. in Manes xi).[4] This passionate, personal response—not a well-reasoned ethical stance—represents for Abbey's characters, as for the radicals of Earth First!, the rationale behind the commitment to active sabotage, an action agenda that Earth First! cannot officially condone despite its predilections and encouragements.

Models for Monkeywrenching

Earth First! came to the attention of the public for the first time in a rally held at the controversial Glen Canyon Dam on 21 March 1981:

> On that morning, seventy-five members of Earth First! lined the walkway of the Colorado River Bridge 700 feet above the once free river and watched five compatriots at work with an awkward black bundle on the massive dam just upstream. Those on the bridge carried placards reading "Damn Watt, Not Rivers," "Free the Colorado," and "Let It Flow." The four men and one woman on the dam attached ropes to a grill on the dam, shouted "Earth First!" and let 300 feet of black plastic unfurl down the side of the dam, creating the impression of a growing crack. Those on the bridge returned the cheer. (Foreman, "Earth First!" 42)

As the organizers were being questioned by the local sheriff, the crowd of demonstrators were addressed by Edward Abbey, who had mythologized the Glen Canyon Dam as the big symbol of all that blocked freedom in the interests of civilized progress. *The Monkey Wrench Gang* had featured a group of characters who shared the ambition of destroying the dam and freeing the Colorado. On the day that Earth First! "cracked" Glen Canyon Dam in their act of guerilla theatre inspired by his novel, Abbey looked beyond symbolic ecotage toward the possibility of the real thing: "Oppose. Oppose the destruction of our homeland by these alien forces from Houston, Tokyo, Manhattan, Washington, D.C., and the Pentagon. And if opposition is not enough, we must resist. And if resistance is not enough, then subvert" (qtd. in Foreman, "Earth First!" 42). The FBI was impressed enough with the performance to dust the black plastic banner for fingerprints, according to "reliable reports" (Foreman, "Earth First!" 42).[5]

Earth First! drew upon Abbey's inspiration not only in its theatrical displays, but also in condoning and perpetrating actual sabotage. According to one report, soon after Mike Roselle, one of the founders of Earth First!, read about the fictional Doc Sarvis' exploits on the highways around Albuquerque in *The Monkey Wrench Gang*, "billboards around Yellowstone"—where Roselle made his home—

"started falling down" (Kane 100). Then in another apparent effort to live out Oscar Wilde's dictum that life imitates art, Earth First!er Howie Wolke followed the lead of Abbey's character George Hayduke and pulled up for the third time the survey stakes of a Chevron Oil Company road-building project in Bridger-Teton National Forest, an action for which he was arrested "at hatchet-point" by a Chevron employee and taken promptly to jail (Malanowski 568).

In 1985, Roselle and other ecoteurs assisted Foreman in the production of a manual on how to monkeywrench and get away with it. *Ecodefense: A Field Guide to Monkeywrenching* gives specific directions for many of the actions first suggested in Abbey's novel: how to use spikes for giving flats to logging trucks or off-road vehicles; how to burn or "revise" billboards; and how to use tools, fire, and water to disable or destroy heavy machinery. (The "classic act of monkeywrenching," pouring Karo syrup into the gas tank of a bulldozer is, thanks to Abbey's novel, the "best-known technique"; but it doesn't really work, according to *Ecodefense* [65]). Abbey wrote the "Forward!" [*sic*] to the manual and endorsed many of the techniques described in print for the first time (though not unknown to monkeywrenchers like Abbey's Aunt Emma, who, he says "has been enjoying the pleasant exercise" of tree-spiking "for years" [*Ecodefense* 5]).

Abbey took considerable steps toward providing such a manual within the novel itself. At the risk of creating wooden dialogue ("How do you get that oil pan loose?" "Well, I'll tell you. . . .") and with little concern over clogging his narrative, Abbey included expansive passages of instrumentally oriented prose, such as this prelude to the gang's demolition of an automated railroad:

> First, they cut the fence. Then they dug out the rock ballast from beneath the crosstie nearest the bridge, on the side of the train's scheduled approach. When a hole was cleared the size of an apple box [*sic*]. Hayduke consulted his demolition card (GTA 5-10-9), handy little item, pocket-sized, sealed in plastic, which he liberated from Special Forces during his previous career [in the U.S. Army]. He reviewed the formula: one kilogram equals 2.20 pounds; we want there charges [*sic*] 1.25 kilograms each, let's say three pounds each charge, to be on the safe side.

> "Okay, Seldom," he says, "that excavation's big enough; you dig another five ties down. I'll place the charge."
>
> Hayduke steps off the railway, back to the sealed boxes waiting on the dune. He rips open the first case—Du Pont Straight, 60 percent nitroglycerin, velocity 18,200 feet per second, quick shattering action. He removes six cartridges, tube-shaped sticks eight inches long, eight ounces heavy, wrapped in parafined paper. He makes the primer by punching a hole in one cartridge with the handle (nonsparking) of his crimping tool, inserting a blasting cap (electrical) into the hole, and knotting the cap's leg wires. Next he tapes the six sticks together in a bundle, the primed cartridge in the center. The charge is ready. He sets it respectfully in the hole under the first crosstie, attaches a connecting wire to the leg wires (all wires insulated) and replaces the ballast, covering, concealing, and tamping the charge. Only the wires are exposed, coiled in their red and yellow jackets, shining on the railway bed. He tucks them under the rail for the time being, where only an observer on foot would be likely to see them. (*Monkey Wrench Gang* 176–77)

The typos in such passages (those indicated by our addition of *sic*, for example) are a kind of index of the impatience an ordinary reader would feel for such detail—readers like copyeditors and proofreaders who are supposed to correct such things. Except in the passages on monkeywrenching techniques and field mapping, this kind of indulgence in realism is absent in the novel, which for the most part, tends toward "hip" expressionism and mock epic descriptions, such as this: "Twenty fathoms under in a milky green light, the spectral cabins, the skeleton cottonwoods, the ghostly gas pumps of Hite, Utah, glow dimly through the underwater mist, outlines and edges softened by the cumulative blur of slowly settling silt. Hite has been submerged by Lake Powell for many years now . . ." (*Monkey Wrench Gang* 34). Or consider this: "Those doomed dinosaurs of iron, waiting patiently through the remainder of the night for buggering morning's rosy-fingered denouement. The agony of cylinder rings, jammed by a swollen piston, may be like other modes of sodomy a crime against nature in the eyes of *deus ex machina*; who can say?" (89). And this description trends from the mock heroic to the utterly ridiculous: "Smith . . . was not aware of how comic and heroic he looked, the

Colorado man, long and lean and brown as the river used to be, leaning on his oar, squinting in the sun, the strong and uncorrupted teeth shining in the customary grin, the macho bulge at the fly of his ancient Levi's, the big ears out and alert. Rapids closing in" (56). Finally, we note this sample of expressivist scenery: "Obscure and ambivalent gloom of dawn. Sky an unbroken mass of violent clouds, immanent with storm" (352)—a natural state mirroring the danger of the heroes, in the manner of the nineteenth-century American romantics. No wonder that one journalist compared *The Monkey Wrench Gang* to an "adventure comic book" (Malanowski 568). But the sections on operations are more like an army training manual with caricatures and photographs to make the reading easy for undereducated G.I.s (the precise graphic techniques used in Foreman's *Ecodefense*).

Abbey could hardly have been surprised, therefore, when Earth First! and other ecoteurs read his novel both as an instruction manual and an incitement to action. The interrelation of fictional and historical actions surrounding the practice of monkeywrenching is indeed complex, with influences doubling back upon influences, and with "life" and "art" ever exchanging places. Abbey was himself writing under the sign of the infamous Ned Ludd, the worker who in 1779 smashed machinery in Leicestershire in protest of the mechanization of labor. *The Monkey Wrench Gang* is dedicated to Ludd and bears a satiric epigraph from Byron: "Down with all kings but King Ludd." Abbey may have also been aware of the work of more recent Luddites—such as "the Fox," an ecoteur who, in the early 1970s, plugged industrial drains, collected polluting effluents, and dumped them on the carpets of corporate offices in the Chicago area (Petersen 22). Earth First! claims that there is much more ecotage out in the world than ever gets reported, since the victimized companies and government bureaus do not want to inspire other potential monkeywrenchers or give them ideas about how to proceed. Abbey also hints that the practice is widespread. During one of their bulldozer-burning campaigns, the characters in *The Monkey Wrench Gang* are surprised to meet another ecoteur about the same business; and in the spiritual climax of the novel, when George Hayduke compliments Doc Sarvis on the "job" he has done on a new bridge across the Colorado River,

Doc confesses that he has had nothing to do with the massive explosion timed perfectly to correspond with the ribbon-cutting ceremonies. The copyright page of the novel offers this "disclaimer": "This book, though fictional in form, is based strictly on historical fact. Everything in it is real and actually happened. And it all began just one year from today."

Motives for Monkeywrenching

Abbey's utopian efforts to write the future and his encouragements to sabotage are matched in historical significance and literary intensity by his exploration of the motives for monkeywrenching. For Abbey, as for other existentialists and romantic individualists in the mold of Thoreau, Whitman, and the beat poets of the 1950s, radicalism arises most directly from personal experience, not from ideology. If something in the life-style or language of the cowboy, the scientist, the construction worker, the citizen, the soldier, the hippie, the Marxist, the environmentalist rings true, then no ideological scruple would be able to keep it out of his expressivist life and literary corpus. Abbey's writing thus coincides with that of the deep ecologists, who suggest that their work is more a form of personal seeking than a systematic philosophy (Devall 5–6); and though Abbey is one of the foremost eccentrics of American literary culture, his depiction of the genesis of radicalism provides a rather close parallel to the public experience of environmentalism as interpreted by Samuel Hays, the American search for an undefiled nature ("environmental amenities") based less in ideology and more in wide-ranging perceptions about the threats of industrial degradation (*Beauty* 246–48). Though Abbey never directly relates the adventures of the monkey wrench gang to the needs of ordinary citizens, neither does he take an ironic or superior attitude toward the public at large, the kind of attitude we have detected in the work of scientific journalists and environmental "experts."

For a writer as acerbic as Abbey, this indulgence of the general public amounts nearly to an appeal. He reserves his satire for those people with direct and immediate connections with the perpetrators of environmental destruction. He attacks the "oil companies, power companies, coal companies, road builders and land developers"—the traditional enemies of reform environmentalism—and because of their compliance with industry and their inability to control developmentalist fervor, he attacks the government bureaus, especially "that limber reed that supple straw that trembling twig the U.S. Department of the Interior" (*Monkey Wrench Gang* 143). But he leaves the people out of it. Ordinary Americans stand on the margins of the action in *The Monkey Wrench Gang,* passive onlookers who watch in wary curiosity as the jack Mormon "Seldom Seen" Smith prays for an earthquake to destroy Glen Canyon Dam, or passive victims who go into surgery with Doc Sarvis, their lungs destroyed by the bad air of Albuquerque.

Rather than seeking to discover links between radical ecology and classical leftist ideologies, such as Marxism (Parsons) or anarchism (Bookchin), discourses in which he had read deeply, Abbey developed instead an informal framework for action based loosely on a romantic allegory—the human being versus the machine. In pursuing this theme, he rebels against the instrumentation, the rationalization, the bureaucratization of modern life as first described in Weber's original portrait of the "iron cage" of instrumental rationality and as developed later in the major work of Lewis Mumford. Abbey would argue that, since Weber's time, the capacity and the range of the iron cage have increased to the point of encompassing the lives and minds of all living beings. In his introduction to Foreman's manual *Ecodefense,* Abbey writes:

> Such is the nature and structure of the megamachine (in Lewis Mumford's term) which is now attacking the American wilderness . . . , the primordial homeland of all living creatures including the human, and the present final dwelling place of such noble beings as the grizzly bear, the mountain lion, the eagle and the condor, the moose and the elk and the pronghorn antelope, the redwood tree, the yellowpine, the bristlecone pine, even the aspen, and yes, why not say it?, the streams, waterfalls, rivers, the very bedrock itself of our hills, canyons, deserts, mountains (4).

The heroes of *The Monkey Wrench Gang* are set against a force of global proportions. The nature of their very actions—the form of their rebellion, the monkeywrenching itself—is determined by the limits that "the machine" sets. Even their ability to think in animal, organic terms is threatened—hence the urgency of their mission: to act while the mind is free. Consider, for example, their various states of imagination as they encounter a strip mine operation. Two of the heroes react with a tragically limited imaginative capability: Bonnie Abzug responds with images produced by the machine of the mass media, the stock images of science fiction; and Seldom Seen Smith merely reflects instrumentally on the size and similarity of the operation as compared to others he has seen: "Their view from the knoll would be difficult to describe in any known terrestrial language. Bonnie thought of something like a Martian invasion, the 'War of the Worlds.' Captain Smith was reminded of Kennecott's open-pit mine ("world's largest") near Magna, Utah" (*Monkey Wrench Gang* 159). Only Doc Sarvis, twenty years older than the others, with broader experiences in life and literature, is able to size up the monster in the terms of the organic imagination, drawing upon the Scandinavian myth of the Kraken, the creature that dwells at the bottom of the sea, rising only to herald the end of the world:

> Dr. Sarvis thought of the plain of fire and of the oligarchs and oligopoly beyond: Peabody Coal only one arm of Kennecott Steel intertwined in incestuous embrace with the Pentagon, TVA, Standard Oil, General Dynamics, Dutch Shell, I. G. Farben-industrie; the whole conglomerated cartel spread out upon half the planet Earth like a global Kraken, pan-tentacled, wall-eyed and parrot-beaked, its brain a bank of computer data centers, its blood the flow of money, its heart a radioactive dynamo, its language the technetronic monologue of number imprinted on magnetic tape. (*Monkey Wrench Gang* 159)

Against the monster, the protagonists are armed not only with their insight that the machine requires human compliance—the soft organic parts drive and control the hard metal—but also, ironically, they have the skills provided by the mechanical culture in which they live. They have only to overcome the division of labor that makes them, as individuals, inadequate to fight the machine. Once they have united through their unique form of communicative action—a mixture of difficult consensus formation, slow-rising but ultimately thorough trust, male-bonding where possible, and sexual sharing to fill the gaps—their talents are formidable; their motives, all the more so, since such motives are far from rare among Americans of their generation. Having survived as a prisoner of war in Vietnam by dreaming of his beloved canyonland, George Hayduke, discovering on his return the work of the Kraken in his own country, turns what he learned from the war machine against itself and fuels his action with pent-up hatred, frustration, drugs, and outright insanity. (Hayduke's "morning mantra" is "Chemicals! Chemicals! I need chemicals!" [*Monkey Wrench Gang* 24].) Seldom Seen Smith—having seen his hometown of Hite, Utah, submerged along with Glen Canyon in the green water of Lake Powell—contributes to the monkey wrench gang his skill as a river guide, learned on numerous forays in the local tourist trade. The bored ex-teenybopper, Bonnie Abzug, brings to monkeywrenching her considerable skill as a liar and sardonic irritant, which she picked up, we must assume, by taking a graduate degree in French literature and by working as a doctor's receptionist. And Doc Sarvis, the surgeon, provides the group's financing by sharing the monetary proceeds—and the analogies—of his work removing cancers from the bodies of the machine's victims. The point is made quite clear—in Abbey's heavy-handed manner—that the monkey-wrenchers are products of the machine, cogs that have gone bad and threaten to wreck the global operation from within. If developmentalism is a cancer on the land, they are the cancers within the cancer. If they have been victimized by what P. M., the author of the ecotopian tract *bolo'bolo*, calls the "Planetary Work Machine," then they will by their work destroy that which has produced them. Their call to battle is "We've got work to do."

The protagonists reflect the author's existential radicalism in their tendency to act upon the slender threads of experience—their overwhelming consciousness of having been victimized and rendered powerless by a machine that controls and destroys their lifeworlds—rather than to drift into the passivity of self-reflective ideology. When, for example, Doc Sarvis and Bonnie Abzug buy a chain saw to help with their work in taking down billboards, the doctor pauses over "the ecological question, whatever that meant, of noise and pollution, the excessive consumption of metal and energy." But he dismisses the "endless ramifications" of the question: " 'no,' the doctor said. 'forget all that. Our duty is to destroy billboards' " (*Monkey Wrench Gang* 42–43).

Abbey romanticizes, of course, and like Earth First!, he is inclined to turn ecotage into comedy. Though the book's main plot climaxes in a rain of gunfire, in which one of the heroes is presumed to be killed, the character emerges nearly unscathed by the end of the novel in a kind of cartoonlike narrative of escape—"You missed." But radicals in history rarely have this luxury. Even while supporting Abbey's followers in Earth First!, the deep ecologist and activist poet Gary Snyder has strong reservations about ecotage: "Any kind of violence in this country will always be outgunned by the government. Their tactic is to lure you into an act of violence, then eliminate you with overkill" (qtd. in Kane 102).

The ecoteurs thus take great risks in their efforts to hurry environmentalist consciousness toward new heights. There are real risks: Workers in sawmills have, according to company reports, already been injured as a result of tree-spiking; and on the other side, two Earth First!ers were hospitalized, then arrested, when a pipe bomb exploded in their car in May 1990. Then there are rhetorical risks. Liberal environmentalists often claim that, though the monkeywrenchers' motives are well-founded, their actions hurt the environmentalist cause overall. "I see no fundamental difference," says Jay Hair of the National Wildlife Federation, "between destroying a river and destroying a bulldozer" (qtd. in Kane 102). Hair misses the point about the limits a mechanistic society puts on action while submitting himself to the narrow range of instrumental action that Foreman and Abbey long to break free of—"commenting on dreary environmental impact statements." The argument that Earth First! muddies the face of environmentalism may be necessary, for reformers like Jay Hair need to maintain their foothold in the Washington establishment; for them to condone violence—against either private property or people—would be the equivalent of negotiating with terrorists. But in many ways, ecotage helps the reform environmentalists both by stalling and frustrating developmentalist progress and by making liberals seem all the more moderate and appealing. Moreover, the liberal argument about the overall image of environmentalism moves in the direction of ecospeak, the effort to divide the discourse of environmental politics into two, and only two, clearly established sides. According to one report, Jay Hair has said that the Earth First! ecoteurs "have no right being considered environmentalists" (Malanowski 569); like many traditional reformers who want to claim environmentalism as their own turf, he finds he must defend it against claims from the right (George Bush) and from the left (Earth First!). Thus, though it may appear that Edward Abbey and Earth First! prolong ecospeak with their oppositional politics and refusal to be appropriated by various ideologies, in fact these radical environmentalists have discovered one way to break the hold of the old dichotomy of developmentalist and environmentalist. The key lies in their insistence on their difference from both the capitalistic exploiters of the earth *and* the genteel liberals of reform environmentalism.

The more serious weakness of radical environmentalism and deep ecology alike—a weakness apparent in the work of Greenpeace and Earth First!—is the overweening negativity of the radical movement as it has developed thus far. Their refusal to think beyond what Manes calls the "unmaking of civilization" opens them to the charge of primitivism.[6] Are the radicals romantically (or sentimentally) calling for a return to simpler times? If so, what golden age provides the model for their ambitions? Writers like Edward Abbey and Dave Foreman seem to dismiss such questions in order to focus their energies on the politics of resistance. In accepting the work of legal reform, however, radicals like Christopher Manes and Mike Roselle have taken the first steps toward creating a positive image of a just and ecologically harmonious way of life. Devall cites the work of deep ecologists now moving in the same direction. We may wonder, however, if their radical anxiety of influence will ultimately keep them from forming hegemonic links with other utopians, with grass-roots citizen groups, and with innovators in global economic politics. Even if productive alliances never materialize, however, the radicals serve as an important catalyst in the emergence of an environmentalist culture; the rhetorical power of their work should not, therefore, be underrated or taken lightly.

The Future of an Environmentalist Culture: Callenbach's *Ecotopia*

In utopian fiction, as in social satire, the poetic aim of creating an image of the world at once different from and suggestive of the reader's world merges with the rhetorical aims of encouraging action and modeling values formation for the reader. But the rhetorical and poetic aims may conflict: the novelist's effort to develop a highly specific blueprint for a new society, for example, can be overtaken by the poetic impulse toward the release of repressed fantasy, thereby increasing audience participation at a deep level while distracting the audience from the author's political project. Or even more likely, the rhetorical aim can overwhelm the poetic, leaving a flat, uncompelling, psychologically and aesthetically unappealing text to support a detailed political program.

Ernest Callenbach's *Ecotopia* is a novel that within the limits imposed by the psychological realism of modern fiction, presents the kind of specific plan for a future community that other radical visionaries (the likes of Edward Abbey and Murray Bookchin) have stopped short of and that more cautious social critics have studiously avoided (see for example, Daly 50; and Illich 15). In addition, Callenbach has discovered an ingenious structure for managing the interplay of poetic and rhetorical aims, developing a plot that revolves around the impressions of a skeptical American journalist, William Weston, who visits the new nation of Ecotopia to report on its curiosities for the mutual benefit of the news-consuming general public and the intelligence-seeking American government. As Weston gradually warms to the Ecotopian lifestyle and ultimately defects, the plot takes on the qualities of an allegory. The story of the hero's conversion to the culture and politics of deep ecology, holistic science, and green technology—in addition to hinting at the author's rhetorical designs upon initially skeptical readers—could well represent the American public's gradual adoption of environmentalist values. The novel is composed of Weston's private diary and his official, public reports sent by wire service back to the United States. Pages from the private and the public record of his journey toward ecological selfhood are interspersed throughout the book so that readers are able to witness how his official mistrust of the Ecotopian economy and way of life is rooted in his personal fears and fantasies of male inadequacy. His gradual conversion to Ecotopian citizenry becomes a mythic rite of passage that models the shifts in consciousness that the author clearly believes are necessary if people are to create political and economic structures in harmony with the needs of the earth. No wonder, then, that the subtitle of *Ecotopia* is *The Novel of Your Future*. Though it fails to achieve an altogether smooth blend of the rhetorical and poetic aims, the novel charts a path for narrative explorations of social ecology, whose goals would be to create, maintain, and extend the institutions of an environmentalist culture.

An Image of Ecotopian Society

Ecotopia is an attempt to answer the question, What if people lived according to the ideals of the environmentalist ethos as it had developed by the mid-1970s? What if people began to shape their living practices by taking seriously the "fusion of resource scarcity economics with holistic biology" to create the kind of "green" community that "ecosophical thinkers" have been working to imagine since the last quarter of the nineteenth century and that deep ecologists have recently proclaimed as a formal normative system (Bramwell xi, 4; Naess and Rothenberg 14–32)?

The fictive nation emerges when the northwest quadrant of the United States—northern California, Oregon, and Washington—secedes from the Union in a bloodless revolution, with the revolutionaries threatening to explode nuclear devices planted by commandos in New York harbor. No one knows if such a device really exists, but the U.S. government takes no chances. (The book was written before the days when President Reagan made a policy of refusing to negotiate with terrorists.) If the methods by which the Ecotopians break free of American control—as well as their general playfulness and freedom with their emotions—suggest the work of Earth First! "ecoteurs," the basic features of the Ecotopian political economy are rather closer to the model of liberal environmentalism proposed by such eco-humanists as Herman Daly and Barry Commoner. (The book's epigraph is a quotation from Commoner's work: "In nature, no organic substance is synthesized unless there is provision for its degradation; recycling is enforced"). Several features of deep ecology, particularly the favored themes of "eco-feminism," also appear, however. The overall result is a fictional model of a "stable-state," ecologically conscious nation run largely by women. The general outlines of the ecotopian political economy thus represent a blend of reform environmentalism, scientific activism, and deep ecology.

Government control is used to ensure the preservation of ecological ideals. Just after secession, the industries (including agriculture) and the transportation system are, for example, nationalized and either dismantled or reformed; displaced workers are absorbed in the construction and maintenance of new, energy efficient public systems of commerce, utility, and transportation; and demands for jobs are offset by the implementation of a twenty-hour work week. The second phase of Ecotopian history witnesses "the radical decentralization of the country's economic life," during which period "the Ecotopians largely [scuttle] their national tax and spending system, and local communities [regain] control over all basic life systems" (Callenbach 79). But, true to the practices of already existing socialisms, the Ecotopians must depend in the short-run on a strong, and at times, a repressive centralized state. William Weston, while still in his unreconstructed state of mind, reports that "the Ecotopian situation has allowed their government to take actions that would be impossible under the checks and balances of our kind of democracy" (23).

A new model of holistic science, one that foregoes the old style of objectivism, guides governmental decisions. Ecotopians are used to buttressing their arguments with appeals to science. Their science, however, appears to have been stripped of the conventions of objectivism. They "spout statistics . . . with reckless abandon" and "have a way of introducing 'social costs' into their calculations which inevitably involves a certain amount of optimistic guesswork" (Callenbach 17). In addition: "Scientists in Ecotopia are forbidden to accept payment or favors from either state or private enterprise for any consultation or advice they offer" (25)—though exactly how they make a living is not clear. One journalist whom Weston meets writes on both politics and science ("not an odd combination here"); he is "skeptical about U.S.

science, which he regards as bureaucratically constipated and wasteful" (39). As with the scientists, so with the journalists: "There is no rule of objectivity, as with our newscasters; Ecotopians in general scorn the idea as a 'bourgeois fetish,' and profess to believe that truth is best served by giving some label to your general position, and then letting fly" (50). Ironically, while neither science nor journalism is restricted to objectivity, advertisements are "limited to mere announcements without impersonated housewives or other consumers, and virtually without adjectives." Although Weston finds it "hard to get excited about a product's specifications-list," he suggests that "the commercials may seem watchable because they are islands of sanity in the welter of viewpoints, personnel, and visual image quality that make up 'normal' Ecotopian TV fare" (50).

Nature mysticism is the de facto national religion. Many Ecotopians are, Weston reports, "sentimental about Indians" (Callenbach 37). Their president, a woman, is "as much a religious leader as a politician" (48). Weston discovers the depth of the Ecotopian religion when he sees his lover, the novel's heroine Marissa, whispering prayers and performing ritual acts in the hollow of a tree. It dawns on him that "this incredible woman is a goddamn druid or something, a tree-worshipper!" (68)

Rhetorical Effects of Fictional Narrative

Rhetorical analysis presses the question: What is gained and what is lost by fitting the programs of reform environmentalism, holistic science, and deep ecology into the framework of the fictional genre? For one thing, environmentalist ideas are no longer presented according to the strictures of what Walter Fisher has called the "rational world paradigm" (3), whose purest manifestation we have seen in the dry analytical format of the environmental impact statement. Nor are the ideas presented in the agonistic rhetoric of reformist polemics, with the pull and tug of the text forcing the reader to take sides. Instead, the novel places ideas into the context of a story and thereby potentially avoids either numbing or alienating the reader. In the form of the popular (and populist) narrative, environmentalism achieves a broader base of appeal and potentially opens into a field of communicative action.

The differences between utopian fiction and other kinds of reformist or radical writing mainly lie in the image of the implied readership for each kind of literature. Writers like Aldo Leopold and Barry Commoner are lecturers; their texts imply a relatively passive, though perhaps intellectually resistant audience, the audience seated in the lecture hall. Greenpeace and Earth First! assume an audience that though potentially active, remains passive in the role of the onlooker, the best image of which is the group of innocent bystanders portrayed in *The Monkey Wrench Gang*. But a novel like *Ecotopia* at least potentially implies an audience of participators, of active fantasists. "Literature," argues the rhetorical theorist Walter J. Ong, "exists in a context of one presence calling to another" (59). The "voice" of the written text may merely "call to" the audience, thereby seeking relatively passive witnesses for an impressive lecture or a dramatic display. Or the voice may "call forth" the audience, inviting participation in the creation of a mutual fantasy or a shared agenda for action.

Ecotopia clearly aspires to the rhetorical power usually ascribed to the evocative mode of calling forth the reader into creative, communicative action. The protagonist William Weston, with his comfortably familiar journalist's voice, invites the skeptical reader to join him on his journey. He wins the reader's initial sympathy with his hardheaded questioning of the Ecotopians and his unreconstructed American materialism. From this base of identification, the reader may stay with Weston as he begins to hear the sense of Ecotopian logic, as he recognizes aspects of a successful technology that outdoes the American version at its own game—efficiency—and finally, as he grows comfortable wearing Ecotopian clothes, as he falls in love with a druid priestess, is wounded (literally and figuratively) in ritualized combat, then healed holistically in an Ecotopian hospital, and ultimately united once and for all with the novel's heroine, the shadow of his own tenderness and receptivity that he has repressed in the personal politics of macho display.

The difficulty comes when, stepping back from our engagement with the text, we readers begin to reflect on the contrasting circumstances of life. There is something of a jolt in the passage from fiction back to historical reality. How are we to know, for example, how many of the traits of Ecotopian life are put forward as serious alternatives to already existing lifestyles and political systems and how many are merely the trappings of satire, such as those used by Swift in *Gulliver's Travels* to shed light upon the social practices of an historical state of affairs? Moreover, how much of the novel's action is devoted to dramatizing the internal state of William Weston? The Ecotopians advocate a kind of free love, for example, and the "nuclear family as we know it is rapidly disappearing" (Callenbach 82). If we take the book literally, reading it as a blueprint for a society reordered on ecological principles, we must see the sexual practices as essential to the functioning of the eco-feminist state. The same is true of the ritualized war games. If we view the novel as instrumentally effective, then we must take seriously the Ecotopian claims that such rituals can really replace war by providing a relatively harmless—and naturally modeled—outlet for the instinctive aggression of human males. But if we allow the novel the ordinary conventions of psychological realism, we may simply look upon these actions as a dramatic means for showing the mental conversion of William Weston, as a rite of passage away from his competitive, "scarce resource" mentality of consumerism in everything from clothes to work to women.

Which aspects of Ecotopian life are presented as actual proposals and which are dramatic devices? When we read the novel under the template of poetic or psychological realism, we tend to ignore it as a rhetorical performance. When we read it under the template of rhetorical realism, we find it lacking and become hypercritical. We notice anachronisms and gaffs in the portrait of Ecotopian environmentalism. The Ecotopians bear a resemblance to the hippies of the late 1960s and early 1970s that is a little too close to capture the historical imagination of the post-Reagan era. Moreover, the inhabitants of Ecotopia are said to be fond of electronic devices and to support a relatively large production of devices that, from our perspective in the 1980s, we know to be produced by industries with increasingly bad reputations as polluters. (The computer and semiconductor industries, for example, were once thought to be "clean" industries but are now frequently stigmatized for air, soil, and water pollution.) And we are left with many provocative questions unanswered. Above all, why are women more likely to be ecologically wise leaders than men; is it because they identify with the ravaged earth under the rule of patriarchy? Has Callenbach literalized what could well be a metaphorical statement of eco-feminism by representing postpatriarchal values in the figure of women priests and political leaders, or is he suggesting that in fact women would be better leaders in this regard?

In final analysis, *Ecotopia*, like the work of Greenpeace and Earth First!, has a great deal of potential power as a consciousness-raising rhetorical performance and has more validity as a critique of existing political and social practice than as a guide to future practice. As such, it is another testimony to the difficulty of thinking our way out of the environmental dilemma, of developing a moral character that harmonizes with our technical knowledge.

The kind of reflection inspired by utopian fiction and by other symbolic actions may nevertheless be crucial in the process by which the environmentalist ethos is formed. The film *China Syndrome*, for example, created a "worst-case scenario" for nuclear accidents that forecasted incidents like the one at Chernobyl so that, by the time the historical events actually occurred, they had come to seem like foregone conclusions—something that could have been avoided. In this way, science fiction can contribute mightily to the radical cause. Since these stories are forthrightly fictitious, they are insulated from the criticism ladled upon environmentalist predictions of real events in the apocalyptic mode. Fiction does not claim the same kind of truth value, so little is lost if its forecasts fail to come true in actual historical circumstances. Under the right conditions, however, such radical visions may well model real revolutionary actions by expressing in language—our chief tool for mediating thought and action—what would otherwise be the stuff of insubstantial fantasy. Walt Whitman, thinking in this vein, described his greatest poetic work as "only a language experiment," but nevertheless insisted that his poems had a more deeply radical potential as "an attempt to give the spirit, the body, the man, new potentialities of speech," potentialities that precede and shape new forms of action (see Killingsworth, *Whitman's Poetry of the Body* 173). The utopian impulse embodied in poetic performance may be a necessary first step in developing the kind of character needed for direct action, or it may provide a language experiment for testing the extremes of potential action. As Richard Rorty has said, "The process of coming to know oneself, confronting one's contingency, tracking one's causes home, is identical to the process of inventing a new language" (*Contingency, Irony, Solidarity* 27).

No reader is likely rush out the door to start the Ecotopian revolution recounted by Callenbach. But such fictions may create a form of environmental consciousness ever more receptive to proposals for new action agendas, such as those outlined in the work of environmental economists like Herman Daly and Lester Brown. These social ecologists demand a yet higher level of audience participation, encouraging readers to compare the details of their own experience with those of other historical peoples or to insert for themselves details conforming to their own histories into a general outline for reform. Above all, the ecological economists urge readers to take action into their own hands, to act creatively now. Approaching their work, we stand at the doorway of a culture infused with environmentalist values.

The response of economists up to now has essentially consisted of dismissing as “utopian” or “irresponsible” those who have focused attention on these symptoms of a crisis in our fundamental relation to the natural world, a relation in which all economic activity is grounded. . . . The standard objection is that any effort to arrest or reverse the process of growth will perpetuate or even worsen existing inequalities, and result in a deterioration in the material condition of those who are already poor. But the idea that growth reduces inequality is a faulty one—statistics show that, on the contrary, the reverse is true. . . . Would it not be more rational to improve the conditions and the quality of life by making more efficient use of available resources, by producing different things differently, by eliminating waste, and by refusing to produce socially those goods which are so expensive that they can never be available to all, or which are so cumbersome or polluting that their costs outweigh their benefits as soon as they become accessible to the majority?

—André Gorz,
Ecology as Politics (13–14)

The hypothesis was that machines can even replace slaves. The evidence shows that, used for this purpose, machines enslave men. Neither a dictatorial proletariat nor a leisure mass can escape the dominion of constantly expanding industrial tools.

—Ivan Illich,
Tools for Conviviality (10)

7

Ecological Economics and the Rhetoric of Sustainability

Think Globally, Act Locally

In a recent column in *Utne Reader,* Walter Truett Anderson writes, "Practically everybody today is some kind of environmentalist. The original movement has diversified into a vast range of organizations, political positions, life-styles, cults, sects, strategies, faiths and fanaticism." Anderson goes on to identify "four distinct wings" of the movement: (1) the *politicos,* whose perspective is more or less the equivalent of what we have been calling "reform environmentalism," the Washington lobbyists and special interest groups like the Sierra Club; (2) the *greens,* which we have been calling deep ecologists or radical environmentalists, those "who want to change society deeply, drastically, and immediately, through protest and massive shifts in lifestyle"; (3) the *grass-roots activists,* those associated with local projects for environmental improvement; and (4) the *globals,* groups like the Worldwatch Institute and the World Resources Institute. Of this last category, Anderson writes, "The really striking difference about the global movement is its emphasis on development. Environmental activism has stressed stopping things—saying no to pollution, technology, new neighbors. But globals are deeply and actively involved in development. . . . Their favorite slogan is 'sustainable development' " (52–53).

We would argue that, though the emphasis on "development" is indeed what differentiates the globals from reform and radical environmentalists, it is rather what links the global movement to grass-roots activism and to the environmentalism emerging in the general public. Though globally oriented economists like Herman Daly and Lester Brown argue for positive, sustainable development, they do not ignore, indeed they have contributed to, the ecologically based critique of standard economics. The globalists themselves would take issue with Anderson's definition of sustainability as "economic *growth* that is not attended by environmental destruction" (53; italics added). Daly, the theorist of "steady-state economics" (an alternative to the gospel of growth), and Brown, the originator of the concept of sustainability, have become powerful critics of high-growth economics and have lambasted the old liberal "illusion of progress" (L. Brown et al., *State of the World 1990,* 3–16). The programs of these new economists call instead for a revision of liberalism toward a social ecology, in which institutions, communities, and individual people promote forms of development rooted in scientific understanding, ecological wisdom, small-scale production, environmentally conscious consumption, and community-based ethics. In developing an interdisciplinary research program that ultimately humanizes and adds an ecological dimension to the theoretical and advisory functions of economics, Daly has created an effective framework for an environmentalist activism that can be applied to personal action as well as to local, regional, national, and international institutions. Taking up where Daly leaves off, and in the same spirit, Lester Brown has formed the Worldwatch Institute, whose primary function is public education, the gathering and disseminating of information on the international economy and its effects on the environment. Not stopping with descriptions, or even with the production of an impressive continuous narrative, the informational system of Worldwatch is devoted also to the shaping of institutional policy and personal action.

In their effort to influence the emerging environmentalist culture, social ecologists cast a broad net, aiming toward a globally effective discourse with universally acceptable values and strong inducements to constructive action. Like Ernest Callenbach, they evoke audience participation at many levels. Taking seriously the popular slogan, "Think Globally; Act Locally," they show particular interest in helping a common readership of nonexperts act in ways that will benefit the global ecosystem. In a paper on "Mobilizing at the Grassroots," collected in *State of the World 1989*, for instance, Alan Durning of the Worldwatch Institute staff tells encouraging stories about small patches of green activity, of local convivial institutions, "an expanding latticework of human organizations that, while varying from place to place in many of the particulars, share basic characteristics":

> The particulars include cooperatives, mothers' clubs, peasant unions, religious groups, savings and credit associations, neighborhood federations, collective work arrangements, tribal networks, and innumerable others. The universals include the capacities to tap local knowledge and resources, to respond to problems rapidly and creatively, and to maintain the institutional flexibility necessary in changing circumstances. In addition, although few groups use the words sustainable development, their agendas in many cases embody its ideal. *They want economic prosperity without sacrificing their health or the prospects for their children.* (L. Brown et al., *State of the World 1989*, 155; italics added)

The appeals of Worldwatch strike a timely note. Grass-roots support for environmentalism in America has shifted in recent years away from an exclusive commitment to resistance—the not-in-my-backyard mentality—to an open commitment to small-scale positive actions like recycling and community education projects that focus on such issues as environmentally conscious shopping, energy conservation in the home, and organic gardening and lawn care. This distinctively middle-class, suburban version of environmentalism was supported by a flurry of publishing activity timed to coincide with the twenty-year anniversary celebration of Earth Day in 1990. New editions of books like *A Sand County Almanac* and *Silent Spring*, as well as new versions of old writings by such authors as Paul Ehrlich and Barry Commoner, were featured at local bookstores across the country. But a new genre of environmentalist writing was more evident than any other—the green how-to manual. Offering 10 or 50 or 101 ways to "save the world" or "live green" or "keep the environment healthy," these books focused on actions that could be practiced by ordinary Americans in their own daily activities. Even *Reader's Digest*, that dependable index of middle-class interest, featured an article on "Simple Ways You Can Help Save the Earth," urging readers not to let "news reports about enormous environmental problems overwhelm and paralyze you": "The 1990s are ushering in a new understanding that government and business can't repair the waste and pollution damage that come from the actions of millions of people," the article announces. "But remember: as much as we are the source of the problem, we are also the beginning of the solution" (EarthWorks 135). Never claiming that grass-roots action will in itself solve global problems, this article and others like it advance, if nothing else, the growth of environmental consciousness by suggesting simple practices in a nonthreatening "Hints from Heloise" format: "tune up your furnace," "subdue your water heater," "monitor your appliances," "don't waste water," "beware toxic wastes," "recycle, recycle, recycle," and "spread the word" (EarthWorks 136–38). Many local utility companies and cooperatives have also begun to offer energy counseling and environmental education services to support consumer conservation efforts.

Reformists and radicals will no doubt look with suspicion upon this turn in the history of environmentalism. They may rightly worry that interest in small-scale actions will function ideologically, blinding the general public to the need for massive shifts in government policy and curtailments of large-scale industrial activity. Consider, for example, a recent television advertisement for plastic garbage bags, a nonbiodegradable and petroleum-based product that has been widely criticized in environmentalist pamphlets for consumers. The ad offers a free guide to "living better in the environment," another contribution to the newly popular genre of the environmentalist how-to manual. The video for the ad shows Boy Scouts filling plastic garbage bags with aluminum cans to be recycled. Ghosts of Indians in buckskin and dressy feathers smile approvingly in the background. And in an overt attempt to displace radical resistance with small-scale consumer actions, the voice-over says, "Don't get mad, get moving!" Reform and radical environmentalists should perceive the need, therefore, to keep up, perhaps even increase, the intensity of their resistance to developmentalist projects that make insignificant allowances to public environmentalist consciousness as a screen for maintaining business as usual.

Moreover, to sustain and support grass-roots activism with a strong critical framework and an action agenda shaped by solid ecological principles and information, environmentalist organizations must develop clear models for action, models based upon a continuous narrative of developmental projects around the world. For all of their accomplishments in issuing a forceful critique of developmentalism, reform environmentalists, scientific activists, deep ecologists, and environmental radicals have offered few concrete plans to help the public map a reasonable course for future action. That work, along with the instrumental-rhetorical discourse that accompanies it, has been carried forward most successfully by economists with environmentalist sympathies.

Unlike natural scientists, economists have rarely hesitated to adopt a public role that allows them to move from *is* to *ought* in their constructions of reality. They have always maintained a high profile as advisors, especially in well-established modern governments. Until recently, however, both the left and the right wings of the field of economics have shared an interest in the developmentalist perspective. Socialist economics and capitalist economics alike have worked toward the end of maximizing economic growth. Only in the last two decades have alternatives begun to assert a stronger influence. Though hardly the first economists to argue for the need to limit growth—a distinction that must fall to the likes of Thomas Malthus, John Stuart Mill, and John Ruskin, who wrote over a century ago—writers in the last two decades have had enough influence to have attained the status of a school of thought. Indeed, a recent collection of essays has contributed the name "ecological economics" (Martinez-Alier and Schlupmann). Following pioneers like Nicholas Georgescu-Roegen and the members of the Club of Rome in the early 1970s, the British economist E. F. Schumacher became the most effective popularizer of the new economics. His *Small Is Beautiful* with its concept of "human-scale technology" has deeply affected the environmentalist movement. To the formulations of Schumacher, Herman Daly has added depth and technical detail in his concept of steady-state economics.

Daly's program has the power of a technical argument, but it departs in so many ways from current economic models that it has the feel of a utopian discourse. Daly brings a flare for language (comparable to Rachel Carson's) and a skill for formal argument (comparable to Barry Commoner's) to the sharp criticisms and the radical innovations he recommends. A close analysis of his rhetoric reveals an implied audience of present reformers and future citizens of an environmentalist culture, to whom he offers a set of logical and ethical tools for clearing out the ecologically inefficient, growth-oriented economy and for building up a steady-state economy based on a scientific understanding of physical forces and relations.

Utopian Rhetoric in Daly's Steady-State Economics

From our perspective, Daly is distinguished as a utopian thinker with a flare for both technical and moral argument; in his work we discover a hearty mixture of reason and character. His *Steady-State Economics: The Economics of Biophysical Equilibrium and Moral Growth* (1977) builds a logical appeal through many affinities with science-based environmentalism and builds an ethical appeal that is founded on an ecological interpretation of the Judeo-Christian concept of stewardship. This book, which we analyze in some detail, provides the basic themes and theoretical underpinnings for his more extended study, *For the Common Good* (1989), coauthored with theologian John Cobb.

Means and Ends

Just as Commoner has done in *The Poverty of Power*, Daly draws upon the work of Nicholas Georgescu-Roegen to establish a foundation for his steady-state economics in the first and second laws of thermodynamics:

> The laws of thermodynamics restrict all technologies, man's as well as nature's, and apply to all economic systems whether capitalist, communist, socialist, or fascist. [The first law:] We do not create or destroy (produce or consume) anything in a physical sense—we merely transform or rearrange. And [the second law:] the inevitable cost of arranging greater order in one part of the system (the human economy) is creating a more than offsetting amount of disorder elsewhere (the natural environment). . . . There is a limit to how much disorder can be produced in the rest of the biosphere without inhibiting its ability to support the human subsystem. We must stop talking about free and inexhaustible gifts of nature and start talking about the throughput, the entropic flow of matter-energy that is the ultimate cost of maintaining life and wealth. (Daly 24–25)

Also like Commoner, as well as Leopold and Schumacher, Daly understands that scientific theory can lead us only so far in our search for a social system based on ecological wisdom. Ultimately, we must build ethical arguments, like Leopold's land ethic, and we must also understand our spiritual underpinnings, much in the way that Schumacher tries to do with his "Buddhist economics." Unlike Commoner, Daly seems more certain of his own ethos and more daring in his projection of it. As early as the mid-1970s, for example, he appears unafraid of the formerly pejorative label attached to ecological reformers, speaking in his preface of his "fellow environmentalists" (Daly x). An even more important difference is that, while Commoner *assumes* many liberal ethical values, using science as his primary base of authority and accepting the role normally assigned to the expert commentator, Daly steps out of the traditional restrictions on the role of economist and tries to develop a systematic and concrete set of relations among religion, ethics, economics, technology, and science.

He argues that traditional economics has suffered because it has ignored not only the basic physical laws of thermodynamics, which represent the *ultimate means* of human action, but also the morality of such action, or *ultimate ends*. He lays out a continuum to demonstrate his points and to show how various disciplines of study have traditionally contributed to our understanding of the relationship of means and ends. This scheme is shown in figure 4 (based on Daly 19).

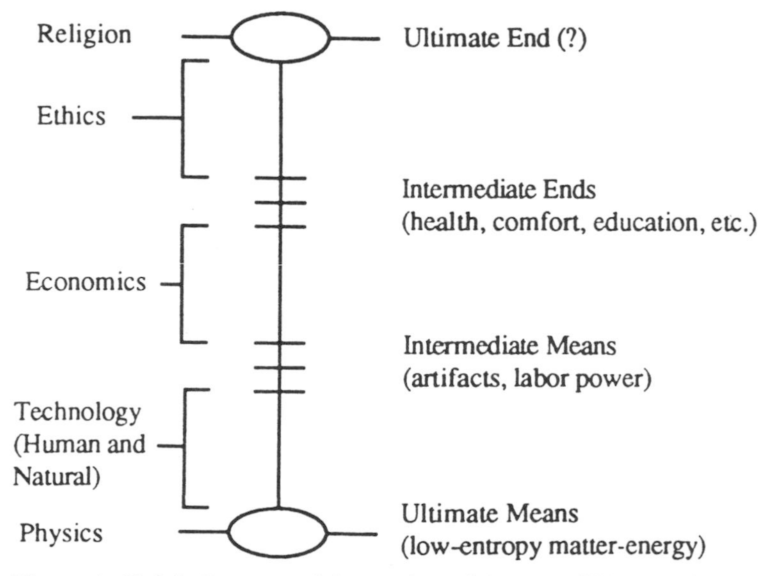

Figure 4. Daly's Concept of the Ends and Means of Human Action

Economics, taking the relation of intermediate ends to intermediate means as its province, has ignored the influence of ultimate ends and ultimate means on the attainment of health, comfort, education, etc., through the use of artifacts and labor power. Ethics, in contrast, has never been entirely cut off from considering ultimate ends, though it may have lost strength in its distance from the physical limitations established by physics in the consideration of ultimate means (hence our concept of *rhetorical stridency*—which results from arguments that in chasing high-minded ends, sacrifice a clear technical understanding of means). Likewise, technology, deriving power and understanding from a close connection with physics, may nevertheless have lost direction because of its distance from a careful consideration of ultimate ends (hence our concept of *rhetorical reductiveness*—which results when good technical discussions of means are uninformed by a clear sense of ethical ends). Ecologically responsible economics—as well as responsible and responsive religion, ethics, technology, and physics—must now pool their knowledges in an effort to create a wisely and efficiently constructed steady-state system of human action.

Daly uses his reflections on ultimate ends and ethics as a rhetorical and logical base for his advice on future policy. Essentially he seeks not only to transform economic practices but to convert the human ethos from a "Faustian" character to an ideal of stewardship, which has roots in both Christian and Buddhist theology. The view of "man as [a] potentially infallible creator seeking salvation in the perfection of his creations," Daly argues, "leads to cosmic vandalism." This Faustian ethos, far from being in line with true science, which Daly wishes to incorporate into his own ethos, is in fact a trivializing, if not a falsifying, view of science's power. Echoing Rachel Carson's criticism of "Neanderthal science," Daly writes of power-oriented science that is untouched by the humility of the steward: "It is the view not of great scientists but of the third-rate devotees of modern scientism." Like the false prophets castigated in the Bible, these devotees of scientism swarm over the earth; like the demons expelled by Jesus, their "numbers are legion" (Daly 26). The steady-state economy "threatens the Faustian convenant with Big Science and High Technology and forces the more humble view that not all things are possible through technology—that the big problems of overpopulation and overconsumption have no technical fixes, but only difficult moral solutions" (39).

Economics must look to ultimate ends and ultimate means as a way of arriving at the humbleness required in living according to ecological wisdom: "The world cannot stand another decade of narrow economists who have never thought about ultimate means or the Ultimate End, who are unable to define either entropy or a sacrament, yet behave as if there were no such thing as entropy and as if there were nothing sacred except growth" (Daly 39). Daly thus requires a double faith in the laws of science and the laws of God in a critique of traditional economics that smacks strongly of the Christian critique of materialism. And yet, he strives to create a nonpartisan image of life lived in compliance with the Ultimate End, an image which, though particularly indebted to the Judeo-Christian and Buddhist traditions, represents a series of common threads present in most of the great world religions, and which also does not fail to ignore the demands described by the physics of ultimate means. From this image he derives a set of "moral first principles": "some concept of *enoughness, stewardship, humility,* and *holism*" (47; italics added).

Toward an Action Agenda and an Effective Political Rhetoric

To realize his moral principles, Daly instrumentally sets forth a concrete program of action for steady-state economics, but he demurs when it comes to providing too many specifics for his plan. He argues that to offer a fully developed utopian plan is a "waste of time" at this stage of history. The key, as he sees it, is to say enough without saying too much: "Drawing blueprints for future societies is a favorite pastime of intellectuals and dreamers, and it is often dismissed as a waste of time. Detailed blueprints no doubt are a waste of time judged by the likelihood that future people will precisely follow their specific impositions. But a general outline or image of a desirable future is an absolute logical necessity for any kind of policy that is not a mere repetition of past practices" (50). Daly thus seeks to provide a "general outline" for three kinds of institutions that, in his estimate, are essential in attaining a steady-state economy. They are: 1) an institution for stabilizing population (transferable birth licenses); 2) an institution for stabilizing the stock of physical artifacts and keeping throughput below ecological limits (depletion quotas auctioned by the government); and 3) a distributist institution limiting the degree of inequality in the distribution of constant stocks among the constant population (maximum limits to personal income and a maximum limit to personal wealth) (53).

Like Commoner, Daly follows the typical environmentalist strategy in looking to government regulation as a corrective for the failures of private enterprise. But unlike Commoner, whose critique of capitalism is utterly devastating and whose recommendation of socialism suggests the adoption of a preformed blueprint, Daly tries to "build on the existing bases of the price system and private property" (Daly 51); his institutions are "fundamentally conservative" (51); "based on respectable conservative institutions: private property and the free market" (70); but are reformatory in that they "are extended to areas not formerly included" in programs of social control: "control of aggregate births and control of aggregate throughput" (51). They thus "provide the necessary social control with a minimum sacrifice of personal freedom, to provide macrostability while allowing for microvariability, to combine the macrostatic with the microdynamic" (51). They differ from an all-out socialism that would nationalize all enterprise and seek to control both micro- and macro-economic processes. Daly argues with both Commoner—who, he says, "leaps to the conclusion that socialism is the only answer"—and Karl Marx, who in Daly's view is not radical enough anymore: "Marx sees capitalists exploiting the soil as well as the laborer. Our analysis sees capital and labor maintaining an uneasy alliance by shifting the exploitation to the soil and other natural resources. It follows that if some institution were to play the role of the landlord class and raise resource prices, the labor-capital conflict would again become severe; hence the radical implications of the ecological crisis and the need for some distributist institution" (111).

The object of Daly's critique is therefore clear: the economics of those people who mindlessly exploit the earth for their own short-term benefit—whether they are capitalists or laborers. But to whom is Daly offering his suggestions for radical reform? Who is the primary audience for this book?

His rhetoric is not designed, we would argue, to convert those whom he criticizes; his critique is too sharply worded, not at all the product of an openhanded rhetoric. Instead, his book appears to be directed to at least two readerships that might consider joining him in his critical practice. One would certainly be his peers in academic economics, to whom chapters like "Efficiency in the Steady-State Economy" seem to be addressed. These portions of the book accept the usual disciplinary limits on argument and thereby retain the professionally certified ethos and logos of the economist. Running a parallel course in the book's structure and dominating other chapters is a thread of argument more carefully calculated to appeal to nonspecialists. More particularly, Daly's book appears to be addressed to the "fellow environmentalists" mentioned in the preface. Consider the following passage, for example, in which an analogy (the economic system compared to a boat) signals the effort to appeal to a nonspecialist (non-economist) audience:

> The internalization of externalities is a good strategy for fine-tuning the allocation of resources by making relative prices better measures of relative marginal social costs. But it does not enable the market to set its own absolute boundaries with the large ecosystem. To give an analogy: proper allocation arranges the weight in a boat optimally, so as to maximize the load that can be carried. But there is still an absolute limit to how much weight a boat can carry, even optimally arranged. The price system can spread the weight evenly, but unless it is supplemented by an external absolute limit, it will just keep on spreading the increasing weight evenly until the evenly loaded boat sinks. No doubt the boat would sink evenly, ceteris paribus, but that is less comforting to the average citizen than to the neoclassical economist. (Daly 69)

The last sentence, in appealing to the common sense of the "average citizen," creates an ironic distance between the reader (whether a nonspecialist "average citizen" or a potential convert from the the field of economics) and the "neoclassical economist," who is pictured as having lost touch with the ordinary realities of life.

By addressing the double audience of fellow environmentalists and potential converts in his own discipline, Daly is able to avoid the stridency we have encountered in the writings of other environmentalists. He must be ever conscious of his fellow economists' demands for effective logos, on the one hand, and of his fellow environmentalists' demands for a new ethos on the other hand. He is thus quite willing to moralize, and even sloganize, economic problems; for example, he says, "Future progress simply must be made in terms of things that really count rather than the things that are merely countable" (Daly 75). This tendency, however, is tempered by his attention to practical and technical matters. We need more than a simple change of character or heart, he argues. So "institutional changes are necessary but insufficient. Moral growth is also necessary but insufficient. Both together are necessary and sufficient, but the institutional changes are relatively minor compared to the required change in values" (75).

After presenting his plan for institutional reform, Daly deals with "two questions [that] must be asked about these proposed institutions for achieving a steady state" (70). The first is technical: Will the plan work? Daly insists that his arguments demonstrate that it will, though he remains open to suggestions from his colleagues in economics and environmentalism: "Let the critic find any remaining flaws; better yet, let him suggest improvements" (70).

The second question has more to do with rhetoric and with the extension of the public environmental movement: Will people accept the institutional changes he proposes? The answer to this is "clearly 'no' in the short run": "The minimum-income side of the distributist institution has some political support in the United States; the maximum income limits will at first be thought un-American"; even if we could agree to set the limit quite high—"let us say, $100,000 per year"—the perception of income size is relative to the individual's own wants and needs in the free market, so someone would feel cheated (Daly 70). Daly's own rhetoric is not yet sufficient to the task of convincing the public to accept his proposals on this matter, nor, he implies, would anyone else's rhetoric do better. Like Commoner, he suggests that history must become the partner of the environmentalist movement. The job of the environmentalist is to prepare for the long run, to make the plans and create the arguments that will lay a rational course that may be implemented when historical and physical circumstances—the very hot summer, the polluted beaches, and the drought of 1988, for example—tip the balance in favor of the environmentalist ethos. Writing in 1989, with coauthor John Cobb, Daly would indeed assert that the historical conditions were ripe for realizing the plan for steady-state economics; the time has passed for the need of strong rhetoric, "wild words," he suggests (borrowing a metaphor from the influential economist John Maynard Keynes); the work of persuasion is now being accomplished by the "wild facts" of history—the thinning of the ozone layer, global warming, acid rain, the spread of toxic pollution to the oceans (Daly and Cobb 1).

Much of *Steady-State Economics* is, at least implicitly, devoted to preparing for the time of the wild facts, teaching the environmentalist audience how to argue against the rear guard of neoclassical economists in the public forum. This objective, particularly clear in the chapter entitled "A Catechism of Growth Fallacies," is accomplished by modeling not only technical arguments but also rhetorical proofs, such as analogies, examples, and definitions. Consider this exploration of political semantics, for example: "The verb 'to grow' has become so overladen with positive value connotations that we have forgotten its first literal dictionary definition, namely, 'to spring up and develop to *maturity*.' Thus the very notion of growth includes some concept of maturity or sufficiency, beyond which point physical accumulation gives way to physical maintenance; that is, growth gives way to a steady state. It is important to remember that 'growth' is not synonymous with 'betterment' " (99).

Daly also cultivates the use of metaphor. He insists, for example, that "the twin sacred cows of property and fertility must be demythologized" (Daly 168). Like Carson, with her implications about the "witchcraft" of pesticide use, Daly speaks of the "black art of econometrics" in his critique of reification in continuous growth economics, and in the same passage, aligns his own discourse with established religion through the use of metaphors adapted from the Bible: "Technology is the rock upon which the growthmen built their church. Since rocks and foundations are concrete entities [a pun on "concrete"?], it is natural that growthmen should begin to endow technology with a certain metaphorical concreteness, speaking of it as a *thing* that grows in *quantity*" (105). In another extended figure of speech, he recalls and extends Leopold's argument that "the land is sick": "Environmental degradation is an iatrogenic disease induced by the economic physicians who attempt to treat the basic sickness of unlimited wants by prescribing unlimited production. We do not cure a treatment-induced disease by increasing the treatment dosage! . . . Physician, heal thyself!" (101). The biblical allusion in the last sentence, the product of Daly's concern with ultimate ends, is among several references to sacred Christian texts, one of which appears in the passage that concludes the book. The prophetic language jars oddly with the economic jargon in the sentence that follows the biblical quotation, indicating that the blending of the moral and technical elements in Daly's ethos is not perhaps as smooth as he might wish: "To stubbornly persist in chasing the expected good at the expense of the offered good would be the greatest possible folly—a folly that the Prophet Isaiah warned about some three millennia ago: 'Why do you spend your money for that which is not bread, and your labor for that which does not satisfy? . . . Incline your ear and come to me; hear, that your soul may live' (Isa. 55.2). Sufficient wealth efficiently maintained and allocated, and equitably distributed—not maximum production—is the proper economic aim" (177).

Daly's rhetoric is sharper and more focused than Commoner's, precisely because a clearer image of his ideal audience emerges. Laying the groundwork for that moment in the long run when environmentalism comes of age as a feasible perspective for social architecture, his work is designed to educate and inform the architects of that future—but without specifying details that would only detract from the general outline of the steady-state economy. Such details would, at any rate, have to be altered to suit the moral and technical needs of the future society as determined by historical circumstances that cannot be foreseen. Indeed, Daly attempts to create a new race of economic advisors and a newly informed and rhetorically prepared cadre of environmentalists.

As we come to the writings of Lester Brown and his colleagues at the Worldwatch Institute, we find arguments designed for and directed to quite a different audience. Writing in the middle and late 1980s, Worldwatch sees its task as historically ripe and indeed urgent in the face of ecological change. Taking for granted the support of Daly's ideal reformers, Worldwatch seeks to influence those who have the power to enforce change in the present—both government and corporate leaders capable of large-scale change and individual citizens capable of small-scale actions. Brown and his colleagues therefore design their writings "to provide a sense of direction for planners and policymakers who are too busy to do all the reading and research needed to make decisions" and to be "used by governments and citizens throughout the world as a sort of touchstone of progress in the search for a sustainable course on the planet" (L. Brown, *Building a Sustainable Society* xii; L. Brown et al., *State of the World* 1990 vii).

Lester Brown, Worldwatch, and the Rhetoric of Sustainability

In his own writings and in his leadership of the Washington-based Worldwatch Institute (founded in 1975), Lester Brown has established the kind of informational and advisory system that the original authors of the National Environmental Policy Act must have had in mind when they came up with the concept of the Environmental Impact Statement. The documents published by Worldwatch follow closely the humanist guidelines for clarity and breadth of appeal that are often set forth, but rarely enough followed, for good technical writing. And even more clearly than Daly's work, Worldwatch literature lays out specific models for action based on impressive but clearly explained technical and moral assessments of world environmental conditions. The institute's notable collection of literature includes regularly published monographs on global environmental economics and politics; *WorldWatch* magazine; and the annual compendium of major Worldwatch papers, the *State of the World* series by Brown et al., begun in 1984. A typical Worldwatch publication is an EIS gone global and carefully sculpted as a rhetorically effective document for a busy readership of policymakers and the general public.

The Art of Accessibility in Worldwatch Publications

Consider as an example a book that amounts to the Worldwatch manifesto, Lester Brown's *Building a Sustainable Society* (1981). The general outline of the book is not unlike Commoner's book-length essays, but the pace of Brown's book is less leisured and the structure is tighter. Part I, "Converging Demands," presents an assessment of current world conditions in agriculture, biological systems (forests, oceanic fisheries, etc.), energy reserves, food needs, and economic and social stresses, developing piece by piece the thesis that high-growth capitalism is not feasible, either ecologically or economically. It is, moreover, dangerous in the short run as well as in the long run. The situation, we are told, is urgent (285). In Part II, "The Path to Sustainability," a plan is laid out for a transition to a new kind of world economic system. Recommendations are presented in thematically arranged chapters on population control, resource preservation, renewable energy use (solar), societal reformation, institutional change, and moral revalutation.

Unlike Commoner and Daly, however, Brown makes every effort to accommodate the busy reader and to develop a presentational plan whose own design reflects the urgency of the situation. *Building a Sustainable Society* encourages fast reading and quick assimilation of information into legislation or other forms of action. The book's structure is characterized by what technical writers call *accessibility*. The author does not flatter himself to think, as humanist authors and academics in general usually do, that the reader will pore over every word in the order it is written. On the contrary, the organization and style are meant to make the text easy to scan; information may be retrieved in a hurry and in the order that the reader desires. The author is thus less interested in controlling or manipulating readers' consciousnesses than in satisfying their informational needs. Accessibility—a potentially powerful effect in the democratic rhetoric of empowerment—is achieved by several techniques, including informative headings; topic sentences; thumbnail essays and narratives; active-voice sentences and strong action verbs; concrete and familiar vocabulary; carefully selected, low-density tables and charts; and other graphical devices to enhance readability. Consider a few examples of these tactics.

Headings are used to break up the blocks of text. They literally give the reader a break every few pages and also reveal the hierarchical principles of the book's organization. The headings are *informative instead of merely structural*; that is, instead of just helping readers find their place in the text—the function of headings like "Introduction," "Results," and "Conclusion"—informative headings provide actual information about what is contained in the section that follows, much like the journalist's headline. The reader can scan the table of contents or the individual chapters and get a fair idea of what a particular chapter says just from reading the headings. Here, for example, is a list of headings from chapter 3, "Biological Systems under Pressure":

Deforesting the Earth
Deep Trouble in Oceanic Fisheries
Grasslands for Three Billion Ruminants
Per Capita Consumption Trends
Future Resource Trends
Oil: The Safety Valve

Topic sentences, carefully written and placed always at the beginning of each paragraph, also help hurried readers, who may well read only the first sentence of every paragraph during scanning. Similarly, the first paragraph of each section, which is, again, the one full paragraph that scanning readers are most likely to peruse, forecasts the major themes to be developed in the section. Consider the first paragraph and the topic sentences of the rest of the paragraphs in a section marked by the relatively uninformative heading "Rereading Ricardo" in *Building a Sustainable Society:*

- The Law of Diminishing Returns was first articulated by David Ricardo, the nineteenth century English economist. He reasoned that at some point additional food could be produced only by extending cultivation onto less fertile land or by applying ever more labor and capital to land. In either case, returns would diminish. (117)
- Initially based on calculations for wheat in the United Kingdom, Ricardo's formulation has a compelling logic. (117)
- As the eighties begin, . . . interest in Ricardo's analysis is reviving. (118)
- While Ricardo's concern with the diminishing quality of new land was initially unfounded, it is now being borne out. (118)
- Just as the quality of new cultivable land has declined, so too efforts to raise land productivity do not pay off as handsomely as they once did. (118)

- Efforts to expand the fish catch represent another clearcut case of diminishing returns. (119)
- With energy as with food, efforts to expand supplies eventually meet with diminishing returns. (120)
- Diminishing returns also govern the mining industry. (120)
- The capacity of the earth's ecosystem to absorb waste also brings diminishing returns. (121)
- Investment in scientific research—long the answer when productivity lagged—may itself be experiencing diminishing returns. (121)
- In retrospect, Ricardo appears to have been ahead of his time. (121)

Since classical times, rhetoricians have recognized that, because of dominant reading habits, information to be emphasized should be placed in first positions—positions at the beginning of chapters, sections, paragraphs, sentences. Brown takes full advantage of this rhetorical rule of thumb in composing the parts and the whole of his books.

Thumbnail essays, like the one on Ricardo's relevance in current economic developments, are placed throughout the book. They are essentially self-contained and thus accommodate the reader interested in only one aspect of the overall treatment of sustainability. This approach contrasts strongly with that of Daly and Commoner, who create book-length essays that require the reader to take in the whole in order to grasp the parts and thus limit the readership to those who have the time and inclination to read several hundred pages. The thumbnail essays generally provide nontechnical explanations that serve as backgrounds and rationales for action programs. Likewise, thumbnail narratives condense what could be a lengthy treatment of historical background into a few readable pages. Consider, for example, these topic sentences from the section headed "Our Petroleum Culture," which tells the story of how oil has transformed global life:

- Oil has left an indelible imprint on virtually every facet of human existence and made our world one that our ancestors would scarcely recognize. (L. Brown, *Building a Sustainable Society* 60)
- To some, the prevailing prices and policies seemed to indicate that rapid sustained economic growth had become part of the natural order. (60)

- Not only did cheap oil underwrite unprecedented economic growth—it also sustained an explosive increase in human population. (60)
- While the rapid growth in oil use greatly enhanced the earth's population-sustaining capacity, it also helped transform the world economy from a collection of essentially independent national economies into a closely knit international economic system. (60)
- Because of its portability and its versatility as a fuel and a feedstock for the chemical industry, oil is widely used in almost all countries, most of which have little or no oil of their own. (61)
- As countries everywhere modernized, the uneven distribution of oil reserves spurred not only trade in oil but also trade in commodities to pay for oil. (61)
- Oil has also been a force in the evolution of the modern international transportation system. (61)
- More specifically, cheap oil led to the evolution of automobile-centered transportation systems in the Western industrial societies and precipitated a desire elsewhere to emulate these systems. (61)
- Automobile manufacturers, the world's largest industry, produced some 100,000 new vehicles each working day during the seventies. (62)
- Along with social change and the integration of the world economy, cheap oil also encouraged the development of "throwaways." (62)

This story sets up, a few pages later, a companion piece on "The Decline of Oil," the grave consequences of which have now been clearly established in a readable narrative that presents not only the commonplace about our dependence upon petroleum and petroleum products but also the less frequently noted structural impact that oil has had on world culture.

Active-voice sentences, strong action verbs, concrete and specific nouns, and a relatively familiar vocabulary are habitually used in Worldwatch publications, not only in narrative passages like the one quoted above, where active language quite naturally supports accounts of action, but also in many passages containing more complex reasoning, as in the following sentences:

- The use of oil in the form of fertilizer and synthetic substitutes for natural materials has served as a safety valve, alleviating the pressure on natural systems. (L. Brown, *Building a Sustainable Society* 56)
- Throughout history, humanity has periodically come up against constraints, but never before has it hit so many in so many places at the same time. (125)
- The transition to renewable energy will endow the global economy with a permanence that coal and oil-based societies lack. (247)

The use of linking verbs, passive voice, and technical terminology does tend to increase in nonnarrative passages, but the use of the more readable forms indicates a strong *preference for narrative prose* wherever it is possible. Most technical prose, as Walter R. Fisher has argued, departs from the narrative structure on its surface but retains in its deep structure its original narrative condition, which is common to all human discourse. In the Worldwatch documents, as in all writings that have accessibility and readability as a goal, rhetorical analysis shows the author's effort to *recover and reveal the original narrative structure in prose passages*, while more schematically presented materials—tables and charts, for example—favor the radical condensation of narrative that requires the readers to unpack the narratives for themselves.

Carefully selected, low-density graphics are essential, then, since they will carry a large informational burden, making up for time lost in more expansive prose passages. Tables or charts appear about every third page in *Building a Sustainable Society* and are carefully calculated to break up passages of prose without incurring informational saturation (see Killingsworth and Gilbertson). These graphics support or extend the arguments carried forward in the written text. In Worldwatch publications, we find none of the combined tables with hierarchies of two and even three items to be compared along the ordinate and abscissa ("nested columns"), such as those we encountered in the scientific papers discussed in chapter 3 or in the EISs discussed in chapter 5. The Worldwatch tables and charts keep the density of the information as low as possible. Moreover, the emphasis on action characteristic of the prose is preserved even in these schematic presentations. The tables and charts, while capitalizing on the summarizing power of the schematic format, maintain the action-oriented rhetoric and the commitment to narrative, often condensing a number of historical actions or proposals for future action into the space of a single page or even a half page. For example, table 1 (based on L. Brown, *Building a Sustainable Society* 286), combines a narrative and comparative framework for presenting two stages of transition to a sustainable society; that is, the narrative movement from an early to a late stage of transition may be compared in several areas of social concern (population, energy consumption, etc.).

Table 1. Early and Advanced Stages of the Transition to a Sustainable Society

	Early Stage	*Advanced Stage*
Population Policy	Introduction of family planning services	Adoption of population stabilization as a social goal
Energy Consumption	Focus on saving energy	Redesign economic system to reduce energy needed
Land Use	Cropland preservation	Land-use planning
Agriculture	Produces food, feed, and fiber	Produces food, feed, fiber, and fuel
Population Distribution	Flow from countryside to cities slows	Population distribution shaped by location of renewable energy supplies
Role of Automobiles	Smaller, more fuel efficient automobiles	More sophisticated planning at community and personal level reduces need for automobiles
Energy Sources	Largely fossil fuels with limited dependence on renewable energy sources	Renewable energy sources dominate
Materials Recycling	Largely voluntary piecemeal programs	Mandatory recycling programs

Source: Worldwatch Institute

The tabular presentation allows the reader to take in a number of actions at a glance and thus to comprehend quickly the kinds of concrete steps needed for achieving the goals of sustainability.

While the tables in the part of *Building a Sustainable Society* devoted to shaping the transition to a sustainable society are used to extend and simplify the presentation of alternative actions, the tables and charts of the background sections are used, much like the denser graphics in scientific papers, to clinch arguments by condensing supporting data and to bolster major themes with visual images. The chart reproduced in figure 5 (based on *Building a Sustainable Society* 73) reinforces the central theme of the rise and fall of oil-based civilization with an image that suggests the shape of a mountain, thus literalizing the concepts of *rise* and *fall*.

Figure 5. World Oil Production Per Capita, 1950–1980, with Projections to 2000

Other graphical devices used to lighten the reader's task in *Building a Sustainable Society* include relatively large type with ample space between lines of type, large margins, and generous white space between sections. The *State of the World* series adds double columns to promote readability and large-print sidebars that repeat key sentences for emphasis (much in the manner of popular magazines) and that add further "breathing spaces" between broken blocks of text. Worldwatch's contract with Norton Publishing Company appears to have resulted in a fruitful match. The effectiveness of the Worldwatch communications has undoubtedly benefited from Norton's considerable experience in textbook publication.

Action-Modeling in Worldwatch Publications

Since Lester Brown's recommendations for action are intentionally transitional and primarily based on factual information derived from recent history, they are not as radical as the more utopian scheme of Daly, but they are more comprehensive and from the perspective of traditional economics, radical enough to be unsettling of the status quo. Nevertheless, they preserve a strong appeal to middle-class values. The mythos of the family, for example, is invoked in the sentence reproduced on the paper cover of *Building a Sustainable Society:* "We have not inherited the earth from our fathers, we are borrowing it from our children." The sentence appeals to the well-known goal of the middle class—giving our children something better than that which we ourselves have known, whether it be a better education, a bigger house, or now, a sustainable society.

The curious optimism of the book is quite likely to appeal not only to the middle-class public but also to politicians who must take a positive case to the voters. If those on the political scene appear to be merely critical—in the fashion of Rachel Carson, the Sierra Club, and the early Barry Commoner—they are likely to be dismissed as morose naysayers, "doom prophets" who have no faith in the American Dream and no concrete plan for "doing it better." (Witness the decline of Jimmy Carter after his famous speech on the American "malaise," which many dismissed as a projection of a personal feeling of depression over his failed policies.) Thus Lester Brown's *Building a Sustainable Society* begins with a cautionary story about the decline of Mayan civilization due to ecologically unwise practices resulting in overpopulation and soil erosion.[1] The situation of our own culture, Brown argues, is much the same: "There are signs that the food problem may unfold during the eighties as dramatically as the energy problem did during the seventies. The parallels are disturbing" (7).

But the chapter ends with upbeat sentences asserting the superiority of contemporary civilization: "What separates us from the Mayans, of course, is our understanding of our environment and our predicament" (8). The resounding first-person plural pronouns indicate the author's enthusiastic participation in contemporary culture and models a rhetoric for the politicians seeking identification with their constituencies. Not only do we have the knowledge, we are even taking steps toward corrective action—"Here and there, the transition to a sustainable society is getting under way" (9)—though we may not be moving fast enough: "Of the many dimensions of the transition to a sustainable society, the most critical is time. . . . Unlike the earlier energy transitions which were relatively leisurely, the shift from oil to sustainable sources of energy must be compressed into the next few decades" (9).

In Worldwatch's *State of the World* series, the accessibility and action-oriented rhetoric are enhanced further by the brevity and thematic focus of the papers included in each collection. Like *Building a Sustainable Society,* these papers tend to give examples of actions contributing to sustainability that are already known and practiced so that their recommendations seem all the more realistic. They say to the reader, "Here's something that has factual status. It has already been done successfully, and you can do it just as well as others have."

In the 1989 paper, "Mobilizing at the Grassroots," for example, Alan Durning documents local organizations that "form a sort of ragtag front line in the worldwide struggle to end poverty and environmental destruction" (L. Brown et al., *State of the World 1989* 154). Grassroots efforts in "developing countries" throughout the world are held up as models and recounted in a one-page action-oriented table that describes the work going on in eleven widely separated nations. A typical entry describes popular efforts in India: "Strong Gandhian self-help tradition promotes social welfare, appropriate technology, and tree planting; local groups number in at least the tens of thousands; independent development organizations estimated at 12,000"; another describes work in Brazil: "Enormous growth in community action since democratization in early eighties: 100,000 Christian Base Communities with 3 million members; 1,300 neighborhood associations in So Paulo; landless peasant groups proliferating; 1,041 independent development organizations" (L. Brown, et al., *State of the World 1989* 157). The table preserves the interest in data and numbers that we have seen at work in environmental impact statements, but it does so without stripping the information of its historical and narrative structure, which is carefully maintained in the use of dates and well-placed allusions (to Gandhi, for example). The rhetorical effect of collecting so many actions in the space of a single page is to create an impression of both richness and simplicity. The actions are richly diverse but have enough in common that they can be presented in a tabular form that requires each action to share a simple criterion with the other actions in the horizontal and vertical columns.

The collected papers of *State of the World* typically follow this pattern of action modeling. For example, the chapter headings of *State of the World 1988*, with the exception of the introduction on "The Earth's Vital Signs," could well be the titles of operations manuals. They are composed of gerund phrases that suggest their commitment to action: "Creating a Sustainable Energy Future," "Raising Energy Efficiency," "Shifting to Renewable Energy," "Reforesting the Earth," "Avoiding a Mass Extinction of Species," "Controlling Toxic Chemicals," "Assessing SDI," "Planning the Global Family" (on population control—with a strong appeal to family values), and as a conclusion, "Reclaiming the Future."

There remains a strong tension, however, between Brown's emphasis on the urgency of the environmental/economic crisis and the optimism with which he sets forth alternatives. One sociology teacher has told us that, despite his own efforts to reinforce the incipient optimism of Worldwatch, a profound pessimism settled over his class when the students read and discussed *State of the World 1988*. The problems are too big to be solved by the measures recommended, the students felt. Progress toward a sustainable society cannot be fast enough. Though this is hardly a systematic study of audience response, it does indicate a possible problem with trying to create an appeal that communicates both a sense of urgency and a plan for action. The insistence on short-run consequences may stimulate a kind of fear and defensiveness that cripples action rather than promoting it.

Moreover, the predictions of short-run failures or successes in current activities may well turn out to be wrong, thus stimulating a dismissal of Worldwatch as a cadre of wolf-criers. Consider the projected decline in world oil production indicated by the downward slope in the line graph reproduced in figure 5 (page 260). The decline is sharper than any historical evidence included on the chart seems to warrant. What if new reserves are found; will this be used as evidence of an overly pessimistic stance on the decline of the oil age?[2]

Brown's studious omission of overt scare tactics like the lurid apocalyptic narratives we have seen in the popular press indicates that he is aware of the need to fit his rhetoric to his rational action agenda. Such rhetorical balance may be the most difficult achievement of the social architect faced with setting forth a rational plan in a dire situation. The success of the Worldwatch documents in achieving a readable discourse marked by a high standard of accessibility—even their relatively successful campaign to seek a rational balance between a discourse that makes people stop and think and a discourse that moves them to action—is no guarantee that this rhetorical difficulty has been solved, that people will not turn their backs on the calls for action because of their fear of the consequences if they fail to act quickly and correctly.

The Prospects of Discourse in an Environmentalist Culture

In conclusion, we may pause over the recent words of Walter Truett Anderson: "Practically everybody today is some kind of environmentalist" (52). Other than the notoriously ephemeral remarks of a journalist, there are many signs of the popular trend toward identifying with what was once a movement of resistance—the environmentalist movement. There are *political signs* like the willingness of the presidential candidates in 1988 to adopt the label "environmentalist." There are *cultural signs* like the rush of schools and universities to add interdisciplinary programs and discipline-centered courses in environmental studies. There are *economic signs* like the eagerness of commercial publishers (and even university presses) to bring out books on environmental topics; the uses of environmentalist themes in advertising; and perhaps most surprising of all, the appearance of "environmentally sensitive investment." By July 1990, one of the leading performers on the mutual fund market in America was an "environmental fund" formed in the preceding February for investors who wanted to avoid contributing to companies that damaged the environment. Such signs point toward the passing of a culture that merely contains within it an environmentalist struggle, a culture whose discourse is ecospeak, and toward the emergence of a culture with environmentalism at its very center.

Even in an environmentalist culture, rhetoric and social politics will continue to play a role as different perspectives are argued on the appropriate forms and levels of public and personal action. But a new discourse will also emerge as the metanarrative or mythology by which the culture carries its values across generations.[3] The continuous narrative of the Worldwatch Institute, especially as realized in the *State of the World* series, "has become something of an institution," as Brown himself says, with justifiable pride (*State of the World 1990* vii). This work may well model the coming discourse, whose purposes reach beyond polemic and policy making and into education, art, social psychology, and myth formation.

The economic concept of sustainability, the central tenet of Worldwatch's metanarrative, will apply equally well to discourse. The favored narratives of a sustainable society will be

- *democratic*, foregoing an elitist retreat from the general public and recognizing the need of all levels of people to have access to reliable information designed to be useful for their particular social goals;
- *open to contributions from diverse sources*, creating new possibilities for hegemonic links, but resisting control by any single perspective or discourse community;
- *action-oriented*, encouraging informed action both by making forthright recommendations and by presenting information in a form reflective of an action context;
- *continuous*, not ceasing to cover important topics, even after the attention of the traditional mass media lags;
- *value-centered*, making no attempt to attain an elusive (or illusive) "objectivity" or neutrality, but nevertheless maintaining a commitment to worthy proofs and following the rules of good evidence;
- *technically competent*, both in style and in content, drawing on the

> best scientific information as it evolves toward factuality, and in the manner of the scientific research paradigm, keeping open to the possibility of changes and shifts in the structure of information.

Such a discourse will not replace the political and normative rhetorics we have studied in this book; indeed it will draw energy and direction from them and in turn will influence their sense of purpose and their understanding of their relationships to other discourses. The continuous narrative of an environmentalist culture will, above all, be the medium through which communicative action is realized and perpetuated.

The problem of democracy becomes the problem of that form of social organization, extending to all the areas and ways of living, in which the powers of individuals shall not be merely released from mechanical external constraint but shall be fed, sustained and directed. Such an organization demands much more of education than general schooling, which without a renewal of the springs of purpose and desire becomes a new mode of mechanization and formalization, as hostile to liberty as ever was governmental constraint. It demands of science much more than external technical application—which again leads to a mechanization of life and results in a new kind of enslavement. It demands that the method of inquiry, of discrimination, of test by verifiable consequences, be naturalized in all the matters, of large and of detailed scope, that arise for judgment.

—John Dewey,
Liberalism and Social Action (31)

Epilogue

The Scientific Activist and the Problem of Openness

May 1990, as part of the twentieth anniversary of Earth Day, we attended a meeting on global warming sponsored by a local environmental action group in Memphis, Tennessee. A panel of three internationally known scientists addressed an audience of over two hundred interested citizens. The information provided was certainly sufficient to arouse the crowd to an awareness of the risks that human technology takes with the earth's atmosphere. We heard that the average overall temperature of our planet has increased by at least one degree Celsius in the last century. This change, we learned, may have come from the normal patterns of climate fluctuation, but there is also the chance that human beings have begun to influence the weather in unprecedented ways; that industrial age technology has overloaded the atmosphere with such a volume of carbon gases that natural processes—the absorption of carbon by the oceans and by green plants—can no longer maintain an efficient state of equilibrium; that the atmosphere, in reaction, may have already become a greenhouse accepting the heat of the sun without releasing it in proper proportions back into space. The destruction of forests by acid rain and advancing industry may have furthered the imbalance, while yet more heat is admitted because the ozone layer of the upper atmosphere has thinned in what is likely a chemical response to yet another industrial product, chloroflourocarbons. If global warming continues at the current rate, the scientists told us, we could face such disasters as rising seas and desertification of prime farm land in the first half of the twenty-first century.

In the estimation of Herman Daly and John Cobb, "wild facts" such as these should have a strong impact on public consciousness even if they are not expressed in the "wild rhetoric" that John Maynard Keynes thought was necessary to stimulate interest in scientifically verified risks (Daly and Cobb 1). Indeed, the sudden newsworthiness of the greenhouse effect in the hot summer of 1988 seemed to bespeak a general awakening to the wildness of climatological data. Citizens eagerly read reports in newspapers and the weekly press, while the United States Senate entertained predictions from scientists on the reality of global warming.

Despite the lift in public interest, however, no global warming policy was created to match the Montreal accord on ozone emissions, which had been formulated by the summer of 1988. The immediate impression of the hottest summer on record faded into adjustments of attitude and expanded efforts in community recycling.

The response of the listeners at the meeting in Memphis reproduced this atmosphere of fading interest. As we glanced around the lecture hall about an hour into the talks, we found that, despite the impressive display of charts and data and the relatively strong warnings about carbon emissions, the eager anticipation of the audience had yielded to widespread yawning and some outright snoozing. Within another half hour, some of the audience—like Whitman, when he heard the learned astonomer—had slipped out to commune with the mystical moist night air. During the question-and-answer session, a local journalist turned unceremoniously to the crowd and announced pointedly that Jonathan Weiner's book *The Next One Hundred Years* had taught her all she needed to know about global warming in an engaging and readable narrative.

Where, then, was the impact of the wild facts? At the time of the meeting, we were completing our research on rhetoric and environmental politics. We should have been able to predict the audience's lagging interest, for what we saw was yet another indication of the gap between science and general human experience. Even though ordinary folks have become a part of what scientists like to call an inadvertent global experiment to determine how much carbon the atmosphere can manage, the nation as a whole lacks enough scientific insight into the early results to do what our scientific mentors seem to be telling us to do—call the whole thing off.

As rhetorical analysts, we have considered the forms that such a warning should take in order to move people to action. We can describe the appropriate genres and strategies. Against such a profound inertia in the public, however, we would be wrong to overstate our optimism about the emergence of an environmentalist culture, implying that the step from altered consciousness to corrective action and reformed institutions is anything but huge. We would also be acting irresponsibly to prescribe definite rhetorical solutions to current discourse problems, solutions that could only be short-term fixes in a social environment of swirling change. In addition to studying language in its relation to action, we have worked in the fields of environmental politics and public education and have seen firsthand how enthusiasm yields to inaction. We would be remiss in closing our book without suggesting the difficulty that writers like those of the Worldwatch Institute face in realizing the virtues of a sustainable rhetoric. The structures of conventional scientific investigation and the mass media, as well as the clanking machinery of government and the passivity of the news consumer, all but ensure that, despite recent shifts in awareness and attitude, people will be slow to act on scientifically generated information, if they act at all. Not to admit the potential for the failure of rhetoric, even rhetoric based on sound information, would be to glorify the arts of the rhetor beyond reason and to underestimate the challenge of rational discourse in a democracy.

In many ways, Daly and Cobb's notion of the wild facts draws upon the Aristotelian view of science, as producing information somehow beyond the need of rhetoric. And since science, according to this view, holds the key to certainty, the path to action lies clearly in the best understanding of nature available. But Daly and Cobb arrive at this view just at a time when neo-Aristotelian scholars are revising their theories of scientific discourse. In two recent books on science and rhetoric, Lawrence Prelli and Alan Gross have come independently to the conclusion that Aristotelian rhetoric must be updated to include science. We must add that scientific information, tempered itself in the fires of rhetoric, must be refined yet again for consumption by a general audience; for if science is not altogether above rhetoric, it is at least rhetorically different from other public discourses. This difference has to be overcome if science is to play a part in the creation of environmental policy. We must learn to see that science is not merely a data base upon which we can rely in making good decisions. It is a view of the world that must be broadened if it is to affect social morality.

Consider, as an obvious example, the cautious attitude that scientists traditionally cultivate toward broad predictions based on limited empirical data. This outlook can prove frustrating both to the general public and to policymakers in need of a clear answer to a question, a definite end to a story. Will we or won't we have to sell our family estate in New Orleans or transfer grain production from Indiana to Manitoba? With its incomparable tolerance for suspense, science is not designed to provide melodramatic closure. The caution derives from the special way scientists understand data. According to Latour's compelling explanation, the scientist's facts are never beyond question; they are only the conclusions of arguments that have been patiently settled by certified procedures within the close confines of the scientific community. "Cold" scientific facts are ever challenged by "warm" scientific disputation. Among the controversies of warm science, we must surely count global climate change.

One rule of the language game in warm science is that the participant should not make a public disclosure of the facts until the scientific community arrives at consensus, until the facts (especially wild facts) have had time to cool. Even then, advice on policy is usually left to engineers and social scientists, for such discourse falls outside the circle of approved scientific conclusions. As a consequence, public interest in science is proverbially difficult to sustain.

The problem for scientists willing to reveal and explain risks to the public is that, while yet maintaining their authority as scientists, they must steer a difficult course between the caution demanded by the scientific community and the closure demanded by the public. In spite of this difficulty, many scientists resist handing over control of public information to writers in the mass media. Often accused of "distorting the facts," reporters are more likely to fail in adequately rendering the tone with which the scientific "source" provides the information; they strip away the proper qualifications and cautions, letting fly with unauthorized conclusions and interpretations.

A good case in point is provided in the work of the climatologist Stephen Schneider, a major contributor to the esoteric field of global climate modeling who, in the manner of Barry Commoner, takes seriously the potential role of natural science in the development of public policy. Frequently quoted as a source on global warming, Schneider has criticized the mass media for overindulging in what he calls "the four D's": drama, disaster, debate, and dichotomy (*Global Warming* 206). After several stories misrepresented his actual views on science and environmental policy, Schneider began to fear for his reputation among his colleagues and was eventually led to popularize his own research as well as the work of other researchers on global climate change ("Both Sides of the Fence" 217). Perfectly aware of what has been called the "struggle between journalists and the scientific establishment for control of information" (Russell 92), he stepped away from the two warring parties and entered the traditional rhetorical realm of polemic, becoming in the process the leading scientific advocate of actions to forestall global warming. In taking the step, he has demonstrated a willingness to risk his professional status—at least up to a point.

In "The Greenhouse Effect: Science and Policy," an article in *Science*, Schneider proves himself an able rhetor, using self-irony to help draw careful distinctions about questions that may variously be categorized as scientific, social scientific, and purely ethical. The arguments are cool and tight, suggesting a man in control of his information and comfortable with his persona. He admits that there are "many remaining scientific uncertainties" (771), and he confronts openly the tentative relation between science and policy. "Whether some amount of scientific uncertainty is 'enough' to justify action or delay it," he insists, "is not a scientific judgment testable by any standard scientific method." "Rather," he says, "it is a personal value choice that depends upon whether one fears more investing present resources as a hedge against potential future change or, alternatively, fears rapid future change without some attempt to slow it down actively to make adaptation to that change easier." Schneider nevertheless maintains, that such a value choice "can only be made efficiently by a society in which those involved in the decision-making process are aware of the nature of the scientific evidence." His main point is that "uncertainties easily reducible in a few years might encourage waiting before implementing policy whereas uncertainties that are unreducible or difficult to reduce might suggest acting sooner" (771).

Writing for his fellow scientists, the primary readership of the journal, as well as the many journalists who use *Science* as a source for their own writing, Schneider displays a keen awareness of the weaknesses in current climate models, perhaps as a foil to James Hansen's claim before the 1988 hearings of the Senate Energy Committee that "the greenhouse effect was '99%' likely to be associated with the recent trends of the instrumental record" ("Greenhouse Effect" 779)—the claim, in other words, that the hot summer of 1988 was almost surely the result of greenhouse warming. Schneider is more cautious. He readily admits that "the complexity of the real world cannot be reproduced in laboratory models," that many of the equations used in the models cannot yet be solved, and that "reliable prediction . . . requires climatic models of greater complexity and expense than are currently available" (774). With these concessions in place, any predictions about local greenhouse effects would seem premature. With self-irony, Schneider speaks of the "array of excuses why observed global temperature trends in the past century and those anticipated by most [Global Climate Models] disagree somewhat" (776), and he asks: "Can society make trillion dollar decisions about global economic developments based on the projections of these admittedly dirty crystal balls?" (775). He even maintains his ironic perspective in admitting the possible benefits for scientists if more research is needed before decisions can be made. More research, he writes, is "appropriate (but self-serving) advice which we scientists . . . somehow always manage to recommend" (778).

As a conclusion, Schneider keeps within the bounds of scientific caution, arguing that "society should pursue those actions that provide widely agreed societal benefits even if the predicted change [in global climate] does not materialize." He writes, "what would be wasted by an energy efficiency strategy? . . . reductions in emissions of fossil fuels, especially coal, will certainly reduce acid rain, limit negative health effects in crowded areas from air pollution, and lower dependence on foreign sources of fuel, especially oil." (778)

Like other articles in *Science*, this well-argued essay becomes an impressive source of information and viewpoints for journalists and policymakers alike. But Schneider refuses to stop here. He also speaks regularly to television audiences, though he admits that "few things are more frustrating than having to condense 500 meaty pages into 500 words that must include two good jokes and one dramatic conclusion" ("Both Sides" 220). In his own writing, he has also opened a direct line of communication with the general public in two books published by the Sierra Club. In his latest, entitled *Global Warming: Are We Entering the Greenhouse Century?*, we can trace the difficulties that the scientist-writer faces in producing prose for a public accustomed to the flamboyance of electronic journalism. Apparently convinced that the general public needs the drama, disaster, debate, and dichotomy that marks the discourse of the news media, Schneider does his best to mold the polemical medium to the needs of the modern consumer.

The book begins with an apocalyptic narrative in the tradition of Rachel Carson's "A Fable for Tomorrow." As we showed in chapter 2, Carson's famous appropriation of the science fiction genre for the purposes of activist rhetoric infuriated her opponents in the environmental debate and alienated a number of her supporters in the scientific community. With Carson's experience in the background, Schneider proceeds in a more circumspect manner. Whereas *Silent Spring* began with a flourish of sentiment—"There once was a town in the heart of America where all life seemed to live in harmony with its surroundings" (Carson 13)—Schneider begins with a warning about his own rhetorical tactics: "No one can know the future, at least not in detail. But enough is known to allow us to fashion plausible scenarios of events that could well take place if current trends and present understanding are even partly true" (*Global Warming* 1). Only after this disclaimer does Schneider launch his story of future life in the "greenhouse century." If Carson's stylized fable drew upon the tradition of ancient wisdom literature, the counterpart offered by Schneider is a parody of a news report, a straightforward narration of sweltering city activities (baseball enduring despite the heat and ozone alerts), browning lawns, health threats to the old and weak, salinity problems in the water supply, threatened shorelines, massive hurricanes, rapid forest dieback, increased risk from toxins, droughts, and regional failures of agriculture. Like Carson, Schneider writes in the past tense both as a means of creating the parody and as a means of bolstering his argument that, though his scenarios probably require a warmer world than we now know, current scientific knowledge suggests that such events are a foregone conclusion if present-day patterns of development continue. We are, he argues, already in the greenhouse century.

Despite this experiment in polemical rhetoric, Schneider carefully guards his authority as a spokesman for science, distancing himself, for example, from nature mystics, whom he dismisses with the label "ecofreaks," and from developmental concerns in business and industry, which he first ignores, then late in the book attacks with slashing irony: "I know it is a very hard line to take," he admits, "but I do not believe that we should prop up dinosaur industries for immediate political convenience when the health of the planet is at stake" (*Global Warming* 75, 264). In constructing the perspective of scientific activism, Schneider is also eager to distinguish research science from the work of hired government experts, for whom he reserves the pejorative term "bureaucrats." "The bureaucratic culture thrives on information," he writes. "Those who live and work in it subsist on a daily diet of cost-benefit studies, flowchart assessments, and so forth." In Schneider's view, this obsession with information serves merely as a means of avoiding the ultimate responsibility for action. Speaking of the recommendations issued by the Toronto conference on global warming, for example, he speculates on the reaction of government officials to the plan: "While environmental protection officials would largely welcome the possibility of more regulatory activities that would expand their influence, they would probably be uneasy with the conference's recommendation for action until further cost-benefit studies showed the impact of the proposed reductions in greenhouse gas emission on various countries and various economic sectors (*Global Warming* 204). Honing his irony on the parodic multiplication of bureaucratic jargon in the last clause—"cost-benefit studies," "impact," "economic sectors"—Schneider hints that the bureaucrats' need to pile up information arises from their anxiety over having an insufficient claim to legitimacy.

Having set his own position apart from that of the bureaucrats, the eco-freaks, and the industrialists, as well as the politicians, the media, and the scientists unwilling to communicate risks to the public, Schneider calls for "an integrated look at all the consequences associated with our actions," an abandonment of "the piecemeal fashion that has dominated policy making in the past," and a "new dimension of creative, integrated solutions" (*Global Warming* 265).

The question that haunts rhetorical analysis is, To whom is Schneider calling? His irony and condescension, his wariness of identifying with some perspective other than that of the research scientist, have very nearly closed off all avenues of appeal. The lines of communication he tries to open to the public may well be blocked by his ironic persona. Though willing to leave aside the cool tone of caution—the certain prelude to a dozing audience—and though willing to sculpt a wild rhetoric attuned to the wild facts of global warming, Schneider is unwilling to take a full step toward *identification* with his audience, the step that is needed in an effective rhetorical exchange, as Kenneth Burke has taught us. Preferring a rhetoric of authority, Schneider retains the kind of perspective that the philosopher Don Ihde has called *Lucretian*: "In his classic *De Rerum Natura*, Lucretius describes the philosophical perspective as one from a high (and, in some translations, 'ivory') tower, from which the woes and movements of humans appear as distant and trivial, like the domain of ants to the standing human. In this distancing—long a metaphor for objectivity—is betrayed the preference for a position both fixed and distant." Updating the language of Lucretius, Ihde develops a new metaphor for the ivory tower that has a striking relevance to the discourse of global climatology: "In a contemporary analogue, we might choose a satellite view. Just as today we have begun to take 'whole earth measurements' via satellite, so the philosopher [or scientist] might prefer this high distance" (Ihde 9). But in Ihde's view, the Lucretian academic is subject like everyone else to a kind of cosmic law of compensation: "*For every revealing transformation there is a simultaneously concealing transformation of the world*" (49). In taking the large view of global climate change, the scientist may well lose sight of the daily grind of workaday life, a life in which, as Idhe effectively shows, ordinary human existence is enfolded with the technologies of transportation, shelter, and production, the very tools and activities that overload the atmosphere with carbon. It is not only "dinosaur industries" but also high-tech industries that contribute to global warming. Schnieder's own computer models rely on chip technology that relies in turn upon the heavy use of CFCs. But such facile assignments of blame mask the ultimate failure of sympathetic vision that eco-humanism requires. The failure of identification implicit in Schneider's text is hardly a one-way street. In many ways, the scientific activist is victimized by a culture that confers upon scientists the loftiest status and then ignores them in their high places of isolation.

In *The Rhetoric of Science,* Alan Gross argues that the "breach between the world of science and our human world is real enough" and that reconciliation is a pressing need. The study of scientific rhetoric, he goes on to say, "is a gesture in the direction of such reconciliation, an argument for the permanent bond that must exist between science and human needs" (20). But to demonstrate that science draws upon the general fund of rhetorical strategies is but one result of rhetorical analysis, an optimistic one at that. It suggests that scientists are not so different from the rest of us. The view offered by rhetorical criticism foregrounds instead the differences between the ethos of science and the perspectives of journalism, government, business, industry, and the general public. These differences drive a wedge in the hope that scientists can bring to the world of environmental policy the kinds of success and power characteristic of their own, more carefully controlled and limited experimental world.

The optimism may only be recovered by a conscious, very nearly utopian, effort to develop open systems of information exchange and policy development. And here we must qualify carefully what we mean by "open." No doubt, just as activist scientists like Stephen Schneider are perfectly willing to make the results of their research accessible to the public at large, a substantial population of readers would be open to receiving the information. Accessibility, though, is not a simple matter of output and input. To get at the complexity of the accessibility issue—and indeed, to take a significant step toward an ecology of information—the founder of "social epistemology,"

Steve Fuller invokes a provocative version of the compensation principle. He argues that the effective dissemination of knowledge is tormented by "the same problems of scarcity that befall other material goods: to make knowledge more available to one place and time is to make it less available to some other place and time." For example, Fuller writes, "To encode quantum mechanics so as to make it accessible to a physicist on the cutting edge of research is, at the same time, to remove it from the first-year physics student or the lay public." He points out that "the crucial epistemological differences occur at the level of the different textual embodiments, since a popularization of quantum mechanics offers the lay reader no more access to the work of the professional physicist than a state-of-the-art physics text offers the professional physicist access to the general cultural issues which interest the lay public." To gain full access to the esoteric world of quantum physics (or global climate modeling) requires an advanced degree in physics earned during a "long period of cloistered study." The meaning of "access," in the usage that Fuller recommends, thus suggests that "A has access to B's work, if A has the capacity to causally influence B's work." Since the layperson who has read nothing but popularizations of physics cannot abide by the rules of evidence used in advanced physics, and since "the physicist, through his professional training in quantum mechanics, cannot inform public opinion," neither has true access to the other's world of action. To be open thus requires a program of personal education, which must have social correlates as well. In Fuller's words, "Placing an advanced quantum mechanics text in every household in the United States is, by itself, unlikely to increase the average American's competence in technically rendered physics. [A] society interested in raising the public's level of competence in physics . . . would have to make a major economic commitment to producing and distributing the texts that would be needed to bridge the epistemic gap implicit in the difference between popular and technical physics texts" (Fuller 271–74).

Crossing the boundaries of discourse communities and creating such gap-filling texts demands the fitting out of new personae, the education—indeed invention—of new kinds of authors and audiences (real and ideal), the conditioning of subject matter to multiple situations, and the adjustment of texts to broad contexts. The requisite blending of genres will be neither an easy task nor an automatic process. The social-epistemological and rhetorical adjustment we face is in every way comparable to the cognitive adjustment required to appreciate Stephen Schneider's joke: "Nowadays everybody is doing something about the weather, but nobody is talking about it" (*Global Warming* 200).

Notes
Works Cited
Index

Notes

Introduction: Rhetoric and the Environmental Dilemma

1. MacIntyre argues that the values of each division of the multi-faceted public are no longer connected with the springs of tradition that inform the arguments, that provide logical and historical coherence to the groups, and that suggest the means by which new arguments may be constructed. In this view, interest groups are separated not only from each other but also from their own founding traditions. MacIntyre's astute description of contemporary problems thus gives way to a conservative set of ethical conclusions that strike us as simplistic and unsound. The contemporary world faces ethical dilemmas unlike those of our historical predecessors, whose worlds were geographically and culturally limited by comparison.

2. Life itself, writes Neil Evernden, "is a continuum that human beings see as a series of categories" (56). The innate human tendency to form hard categories is, according to Evernden, at the root of our inability to think our way out of the environmental dilemma. In his phenomenological view, "the division of the world into subject and object is trivial" (58). Human consciousness enfolds and unifies the field that contains the continuum of human beings and nature. Moreover, the greater our ability to specify unique and multiple positions along the continuum—to provide what we are calling, after Laclau and Moeffe, "subject positions"—the greater chance we have for clear understanding. As Evernden suggests, "our understanding of any phenomenon is generally hampered by a restriction in the number of categories recognized. . . . The greater the range—that is, the finer the distinctions we make within the spectrum—the more subtle our knowledge of the world" (75).

3. We are grateful to Professor Anne Rosenthal of Illinois State University for calling our attention to the work of Laclau and Mouffe.

1. Varieties of Environmentalism: A Genealogy

1. See Hays, *Conservation and the Gospel of Efficiency*. We give a fuller treatment of the political and rhetorical consequences of Pinchot's heritage in chapter 5.

2. As Graber notes, the "caricature" of wilderness purists as " 'backpack snobs,' an 'aristocracy of the physically fit,' who criticize the harmless pleasures of social campers as part of an effort to monopolize scarce resources,"

is an image that "dies hard because it contains an element of truth" (80). Studies of wilderness users have provided a profile that lends some credence to Watt's use of this well-worn stereotype. Indeed, these users tend almost exclusively to be young and well educated. But "although wilderness users are highly educated and hence represent the professional and managerial occupations, they are not a wealthy and leisured elite" (Graber 17). Watt's own rhetorical stance, particularly his championing of the handicapped, would no doubt have seemed ironic to the many media commentators who later criticized Watt's infamous faux pas in his last days in office, his sardonic quip about the selection of members of a coal advisory board—"I have a black, I have a woman, two Jews and a cripple" (Rosenblatt 100).

3. Along similar lines, Evernden writes, "If we encounter nature as natural resources, then we deny it any of the character of worldhood. And we simultaneously deny ourselves access to it as home. It is characterized by space, not place. There is no human involvement and therefore no sense of significance in such a nature" (66). Evernden depends heavily upon the German philosopher Martin Heidegger's conception of *Dasein*—literally "being-there"—which ties the primordial sense of being to a location, a dwelling-place, a homeland, a historical reality. Heidegger's early fascination with the Nazis no doubt was linked to Hitler's romanticizing of the "fatherland." The historian Anna Bramwell has noted that the same link has been attributed to the contemporary German Greens, who inspire among some the fear of a "potential return to small-scale European nationalism" and "the danger of breaching one of the main conventions of Western democracy since the War, the centrality of the Jewish experience under the Nazis." Bramwell insists, "This is not because there is actual anti-semitism among the Greens . . . but because they implicitly turn their backs on so many of the Enlightenment [that is, modern or liberal] ideas: progress, emancipation, growth and utilitarianism" (224). Bramwell calls Heidegger, who is frequently quoted by Green politicians, the "metaphysician of ecologism" (11). Despite the disturbing overtones of Heidegger's language and political associations, the depth of his concept of *Dasein* should neither be slighted nor dismissed on ideological grounds. In a late essay, "Building Dwelling Thinking," written after the destruction of the German homeland in World War II, he points the way beyond the initial conflict of unity versus disunity with nature, dwelling-in versus homelessness: "as soon as man gives *thought* to his homelessness, it is a misery no longer. Rightly considered and kept well in mind, it is the sole summons that calls mortals into their dwelling" (339). *Homelessness*, a word whose literal and figural implications are more grave than "footlooseness," which Schumacher applies to more or less the same problem, is a key social issue of our times. Being has been separated from

its there-ness. The urban interest in environmentalism is likely spurred by the need to relocate the self in its lifeworld.

4. In a similar vein, the philosopher of science Ian Hacking writes: "whenever we find two philosophers who line up exactly opposite on a series of half a dozen points, we know that in fact they agree about almost everything. They share an image of science . . ." (5). Likewise, Berry and his opponents, perhaps even George Bush, share an image of human alienation.

2. The Rhetoric of Scientific Activism

1. Leopold builds his appeal upon what, according to neo-Marxian social criticism, is a common feature of bourgeois thought—the substitution of "nature" for history. In the formulation of Roland Barthes (who extends the analysis of Marx and Engels in *The German Ideology*), myth, the "depoliticized speech" of bourgeois rhetoric, "transforms history into nature": "The world enters language as a dialectical relation between activities, between human actions," but "it comes out of myth as a harmonious display of essences. A conjuring trick has taken place; it has turned reality inside out, it has emptied it of history and filled it with nature" (Barthes, *Mythologies* 129, 142). For a treatment of this rhetorical tactic in the poetry of Walt Whitman and in nineteenth-century social reform literature in America, see Killingsworth, *Whitman's Poetry of the Body*.

2. The rhetoric of "A Fable for Tomorrow" has proved extremely influential among writers who have followed Carson in applying scientific activism to problems involving an extensive environmental threat. In recent years, for example, the apocalyptic narrative has been widely used in writings on global warming due to the greenhouse effect. "No Escape from the Global Greenhouse," an article printed by the British journal *New Scientist* on 12 November 1988, picks up much of the tone of *Silent Spring*, including overtones of magic gone awry—a gothic icon depicting mysterious owls and skeletal human figures rising above an earthly plain is centered in the two-column format, just above the middle of the page—along with a preference for alarmist rhetoric: in the last sentence of the first paragraph, a classic position of emphasis, we read, "Global warming is threatening life on this planet" (Pain 38). The author launches an apocalyptic narrative, which instead of using the past tense to soften the effect of the rhetoric and skillfully to transform it into a fable, in the manner of Carson, speaks confidently and prophetically in the future tense, with little effort to hedge on the certainty of the claims with conditional verbal structures (such as the verbs *may* or *could*). For example, Pain writes, "Other, equally damaging changes will accompany the rise in temperature. Sea level will rise both because water

expands as it heats up and because the polar ice will begin to melt. In many parts of the world, storms will be more frequent and more ferocious; heat waves will be longer and hotter and droughts protracted. Elsewhere, rains will be more severe. Some scientists believe that this year's drought in the US and the recent spate of storms and hurricanes are all signs of the changing climate—and herald worse to come" (Pain 38). Somewhat more circumspect, but just as confident, is the activist work of the atmospheric scientist Stephen Schneider, who begins the first chapter of his book on global warming with an apocalyptic narrative in the tradition of *Silent Spring*. We discuss Schneider's rhetoric in our epilogue.

3. The historian Anna Bramwell looks upon this kind of extension of radical politics with withering irony: "After the failure of the student movement in 1968, some Berkeley activists found a new cause. Marxist criticisms of alienation and reification combined with a Reichean critique of hard, paternalist insensitivity. Rachel Carson's . . . *Silent Spring* had demonstrated the existence of an astonishing degree of pollution of North America's vast and fertile sub-continent. The young radicals now claimed that it was multinational capitalism that was responsible for pollution. It was the positive value behind this socialist criticism that provided a non-party, catch-all popularity among America's affluent middle-class. At last radical socialism could combine with aesthetic values. The urban proletariat was no longer God; it would, indeed, be abolished all together. Political action in the form of campaigning against ecological damage could now morally be carried out from a comfortable suburb. The 'pink-diaper babies' found Marcuse more spiritually appealing than Marx." (225).

4. No wonder that Stephen Schneider, another scientific activist interested in preserving his credentials in mainstream science, carefully distances himself from "ecofreaks (people looking for oneness to nature)" (*Global Warming* 75). Commoner's rhetorical experimentation with mystical themes is surely a great risk to his credibility among scientifically trained readers.

3. Scientific Ecology and the Rhetoric of Distance

1. Kuhn at least allows the occasional intrusion of technological influences; the "crafts," he suggests, have contributed both instruments and facts to the development of the sciences, especially in their most primitive stages (15–16). (Compare the historian Derek de Solla Price's famous assertion that the steam engine has done more for thermodynamics than thermodynamics has done for the steam engine.) For his part, Popper assumes the context of an "open society," without which his version of science would be impossible.

2. Since we did our original research for this chapter, Hinman has followed that path himself, turning from academic research to a job in industry. In

the wake of the disastrous Valdez oil spill, he joined the environmental research division of the Exxon Corporation.

3. Lyotard goes on to sound the death knell of these grand narratives and to announce, as the "postmodern condition," the reemergence of the diverse and heterogeneous discourses of scientific and micronarrative knowledge. We believe that he is premature, and perhaps reactionary, in taking this step. We find more cogent Habermas' assertion that the liberal modernist project with its grand narratives still prevails in the West, that "we remain contemporaries of the young Hegelians" (*Philosophical Discourse of Modernity* 53).

4. Transformations of Scientific Discourse in the News Media

1. A good introduction and reliable annotated bibliographies on the key topics of science journalism are available in *Scientists and Journalists: Reporting Science as News,* edited by Friedman, Dunwoody, and Rogers. See also Goldsmith. The extensive literature produced by media studies specialists tends to give an insider's view of science news, which contrasts a good deal with the (admittedly partial and limited) perspective we offer in our brief treatment here.

2. Hansen earned the contempt of many scientists not because he made this statement, but because he went public with his findings before publishing them on 20 August 1989, in a refereed article (Hansen et al). He not only testified before Congress, but granted several interviews and held press conferences for the benefit of the mass media prior to the official appearance of this publication. Stephen Schneider has been criticized on similar grounds. We discuss his experiences in our epilogue.

3. Compared to our analysis of the stories in *Science,* our treatment of *Time* may seem rather abbreviated. Our assumption is that most readers will be more familiar with popular journalism than with the insider journalism of *Science.* We are counting on our readers' ability to fill in examples from their own reading in the popular press.

5. The Environmental Impact Statement and the Rhetoric of Democracy

1. We do not presume to give, in our limited space and for our limited purposes, a full rendering of the theoretical opposition of instrumental and communicative rationality as developed from Weber through Horkheimer, Adorno, Marcuse and down to Habermas. This theoretical work is important enough for the discourses we analyze, however, that we are compelled to give at least a simple outline of it and to hint at the connection of it to our own understanding of the relation of discourse to other forms of action.

2. Communicative action is the ideal in the primitive democracies de-

scribed by Lewis Mumford: "The spinal principle of democracy is the perception that the traits and needs and interests that all men share have a superior claim to those put forward by any special organization, institution, or group. . . . Democracy, in the sense I here use the term, is necessarily most active in small communities and groups, whose members meet face to face, interact freely as equals, and are known to each other as persons: it is in every respect the precise opposite of the anonymous, de-personalized, mainly invisible forms of mass association, mass communication, mass organization. But as soon as large numbers are involved, democracy must either succumb to external control and centralized direction, or embark on the difficult task of delegating authority to cooperative organization" (*Technics and Human Development* 236).

3. Bruno Latour, in his study of scientific rhetoric, offers a new twist on the concept of instrument that hints strongly at a similar self-objectification among research scientists: "I will call an instrument (or *inscription device*) any set-up, no matter what its size, nature and cost, that provides a visual display of any sort in a scientific text" (68). This definition "does not make presuppositions about what the instrument is made of. It can be a piece of hardware like a telescope, but it can also be made of softer material. A statistical institution that employs hundreds of pollsters, sociologists and computer scientists to gather all sorts of data on the economy is an instrument if it yields inscriptions for papers written in economic journals. . . ." (68–69). Even a single scientist could be an instrument under this definition: "a young primatologist who is watching baboons in the savannah and is equipped only with binoculars, a pencil, and a sheet of white paper may be seen as an instrument if her coding of baboon behaviour is summed up in a graph" (69).

4. For a full discussion of the problem of cognitive saturation in the use of graphics, see Killingsworth and Gilbertson, "How Can Text and Graphics Be Integrated Effectively?" 138; also Turnbull and Baird.

5. In regard to the practice of reification in the environmental impact statements, Hays writes, "NEPA has required agencies to devise ways and means of taking into account intangible or unquantifiable factors. But it would be difficult to identify a case in which this was done. Agencies were prone to deal with the problem, if at all, only by attempting to turn an unquantifiable factor into a quantifiable one by devising some scheme to attach numbers to it. But this only continued the practice of giving less importance to the unquantifiable" (*Beauty* 281). A good nontechnical account of dehumanization in procedures like environmental assessment is given by Barry Commoner in his 1987 "reporter at large" article in *The New Yorker*. In a discussion of air pollution, he writes, "The cost of the necessary controls [anti-pollution devices] is compared with the value of the benefits

(to human health, for example) generated by a reduction in the level of pollution. Such an analysis attempts to give a social aim—the protection of the environment and of public health—dollar dimensions. Transformed into a pseudo-economic value, the social value can then be compared, it is supposed, with other economic factors, like the cost of controls, and some 'cost-effective' balance struck" (63). Friesema and Culhane similarly argue that EISs "usually consider only one social consequence—the economic impact of the project" (343). Commoner warns that the "apparently 'objective' cost-benefit approach" and the related "risk-benefit computations" used in impact assessment are "quickly engulfed in deep moral issues" (63–64). The classic critique of this form of reduction is given by Marx and Engels in *The German Ideology*. Of the commercial revolution in nineteenth-century England, they write: "It destroyed natural growth in general as far as this is possible while labour exists, and resolved all natural relationships into money relationships" (57). For a full exposition of Marx and Engels' argument that capitalism "dehumanizes man and denatures nature," see Parsons (17–19 and throughout). Parsons understands Marxism as a humanizing system, while Callinicos, in an effort to reconcile Marxist philosophy with Anglo-American analytical philosophy, reads it as a realism. Marxism may be viewed as a humanism only so far as it opposes, dialectically, reification, which Marx understood as characteristic of the capitalistic mentality: "In labour all the natural, spiritual and social variety of individual activity is manifested and is variously rewarded, whilst dead capital always shows the same face and is indifferent to the real individual activity" (Marx and Engels 23).

6. As Jorge Larrain notes in his study of the concept of ideology, "the dominated . . . spontaneously formulate their grievances in the language and logic of the dominant class" (157). In a highly suggestive, but as yet unpublished, paper, Professor Valerie Balester of Texas A&M University refers to this type of incomplete digestion of an alien style as "hyperfluency," the stylistic counterpart of the grammatical condition that linguists call "hypercorrection."

7. A giant in the field of limnology, the study of lake ecology, Schindler is the leader of the Experimental Lakes Area research group in Northwestern Ontario. The Canadian government has deeded to this project a number of small lakes. The researchers manipulate the chemistry of these lakes in order to simulate conditions in endangered environments. They systematically add acids, for example, to predict the effects of the continued increase in the acidity of lakes polluted by acid rain. Though deep ecologists will groan at this violence against the earth's body, the results produced by Schindler's experiments have led to significant reforms and environmental improvements. He challenged, for example, the findings of one government commis-

sion which had claimed that the expensive phosphorus control measures demanded by legislation would not significantly improve the eutrophication of Lake Ontario. Schindler's data argued against the commission's results. He and his colleagues were able to convince the government not to repeal the control measures: "Within a few years, there were clear signs of recovery in the lake" (Schindler, "Detecting Ecosystem Responses" 7). Schindler is also a strong critic of computer simulations of environmental conditions—which he claims are not an effective substitute for direct manipulation and observation of the environment in the post-Baconian tradition.

8. On the status of the EIS as a genre, see Miller, "Genre"; and Killingsworth and Steffens.

6. Rhetoric and Action in Ecotopian Discourse

1. There is also strong tension between deep ecologists and *social* ecologists, also known as eco-humanists. See, for example, the acrimonious exchange between eco-humanist Murray Bookchin (*Remaking Society: Pathways to a Green Future*) and deep ecologist Christopher Manes (*Green Rage: Radical Environmentalism and the Unmaking of Civilization*).

2. None of the reports in the news media have touched on Greenpeace's appeal to labor, probably because it violates the premises of ecospeak, particularly the assumption that environmentalists characteristically stand in opposition to job development in community formation. But Greenpeace, in an effort to reach beyond its commitment to the wilderness ethic, has recently taken quite seriously the plight of workers and of the underprivileged people who live in the most environmentally degraded habitats of American inner cities. David Lyons, a journalist from Memphis, Tennessee, told us that during Greenpeace's inland waters campaign against toxic waste dumping, Greenpeace campaigners appeared on the same platform with union organizers working in industries in southern Louisiana. The toxic waste issue is one that cuts across any division between environmental radicals and labor interests, as is the issue of workplace safety.

3. Earth First! militants are especially conscious of dress and other outward signs of establishment identity. "Mainstream environmentalists are out of touch," says Foreman's friend Mike Roselle; "Most of them are in D.C., doing lunch in their designer khakis and working out their retirement bennies" (qtd. in Kane 98). The critics of the group would no doubt dismiss this preoccupation as superficial. Unlike Foreman's crude historicism, however, their sensitivity to signs and rhetorical effects is often quite subtle and their interpretations strikingly cogent. From their early efforts at "guerilla theatre" (Malanowski 568) to Manes' recent borrowings from the radical semiotics of Barthes and Foucault, Earth First! has insisted on radicalizing every detail of environmentalist psychology and sociology.

4. Here Abbey was reiterating a view that he first set forth in the introduction to Foreman's book *Ecodefense*: "If a stranger batters your door down with an axe, threatens your family and yourself with deadly weapons, and proceeds to loot your home of whatever he wants, he is committing what is universally recognized—by law and morality—as a crime. In such a situation the householder has both the right and the obligation to defend himself, his family, and his property by whatever means are necessary. This right and obligation are universally recognized, justified and even praised by all civilized human communities. Self-defense against attack is one of the basic laws not only of human society but of life itself, not only of human life but of all life" (*Ecodefense* 3).

5. Such "reports," though frequently repeated in the news media, should of course give pause to the historian. Lying—or what P. M., the anonymous author of the poststructuralist ecotopian tract *bolo'bolo*, calls "dysinformation"—is a respected practice among saboteurs. If information, like power and control, is what "the Machine" thrives upon, then it too must be subverted. Unlike historians, however, rhetorical analysts who recognize the structural and instrumental connection between fiction and historical events must, with sufficient irony (and humility), admit some sympathy with the perspective of the Earth First! ecoteurs who, according to the journalist Elizabeth Kaufman, have some trouble in "determining at what point the theater ends and real life begins" (119).

6. Of the Green party in Germany, the historian Anna Bramwell remarks: "Green representatives display impressive debating skills when dealing with the kind of politician who will explain to them that they represent marginalism in late capitalist society. But they are less impressive at producing proposals for general environmental improvement" (221). She goes on to say that "their hope of regeneration presupposes a return to primitivism, and thus, whether they wish to enunciate it or not, concomitant anarchy, the burning before the replanting, the cutting down of the dead tree" (248). This critique cannot be taken, however, as sufficient grounds for dismissing radical environmentalist critique. As Barry Commoner has recently pointed out, most improvements in pollution levels in the United States may be attributed to *negative* actions—stopping the use or production of particular technologies and chemicals, or placing heavy restrictions on such use and production ("A Reporter at Large").

7. Ecological Economics and the Rhetoric of Sustainability

1. The rhetorical nature of Brown's use of the analogy to the Mayans is suggested by his failure to note just how controversial this thesis is. Anthropologists are divided over the exact causes of the decline of Mayan

civilization. The topic is still a "live" one, warmly debated, and is considered a mystery in many circles. Ironically, some environmentalists point to the Mayans as a model community in harmony with nature. The use of the Mayan thesis as a frame for *Building a Sustainable Society* is one of many instances in which Worldwatch uses unsettled scientific claims to further rhetorical purposes.

2. The rhetorical danger of wolf-crying, which we mentioned in connection with James Hansen in chapter 4, cannot be overemphasized. To make a wrong prediction is to label oneself as a doom prophet and to cast doubts on one's authority as a scientific advisor. Consider this typical passage from a syndicated column by the conservative journalist Warren T. Brookes: "Fortunately, [Senator Albert] Gore's global warming scenario is scientifically unsustainable conjecture by people like [Paul] Ehrlich, who predicted in 1968 that because of food production limits, 'by 1985 enough millions will have died to reduce the Earth's population to some arbitrarily acceptable level like 1.5 billion people.' Instead, world food production has increased at an even faster rate ever since" (Brookes A4). Brookes uses this mistake to justify labeling Ehrlich as "Stanford University's doomsday biologist" and dismissing Ehrlich's assertion during a television appearance that because of global warming "we are going to see massive extinction . . . lose all of Florida, Washington, D.C., and the Los Angeles basin. . . . We'll be in rising waters with no ark in sight" (qtd. in Brookes A4). The intellectual historian Anna Bramwell has been equally critical of "reputable scientists like the Ehrlichs and Barry Commoner [who] drew attention to pollution that was real enough (however unreal and indeed disproven their prophesies of disaster)" (225), noting that "the apocalyptic visions of twenty years ago (hundreds of millions dead of famine, infertile soil, climatic disaster) have not come true" (227).

3. A full description of the emerging institutions of an environmentalist culture is beyond the scope of this book. Such a work would require, in addition to rhetorical criticism, some serious ethnographic investigation. We are attempting to incorporate this methodological extension in our current work in progress.

Works Cited

Abbey, Edward. "Canyonlands and Compromises." *Voices for the Earth: A Treasury of the Sierra Club Bulletin*. Ed. Ann Gilliam. San Francisco: Sierra Club. 1979: 391–93.

———. *The Monkey Wrench Gang*. New York: Avon, 1975.

Anderson, Walter Truett. "Green Politics Now Come in Four Distinct Shades." *Utne Reader* July–Aug. 1990: 52–53.

Arms, Karen, and Pamela S. Camp. *Biology*. Philadelphia: CBS College, 1982.

Bardach, Eugene, and Lucian Pugliaresi. "The Environmental Impact Statement vs. the Real World." *Public Interest* 49 (1977): 23–38.

Barilli, Renato. *Rhetoric*. Trans. Giuliana Menozzi. Minnneapolis: U of Minnesota P, 1989.

Barthes, Roland. *Mythologies*. Trans. Annette Lavers. New York: Hill, 1974.

Bazerman, Charles. *Shaping Written Knowledge: The Genre and Activity of the Experimental Article in Science*. Madison: U of Wisconsin P, 1988.

Beale, Walter H. A *Pragmatic Theory of Rhetoric*. Carbondale: Southern Illinois UP, 1987.

Begley, Sharon. "The World's Eco-Outlaw?" *Newsweek* 1 May 1989: 70.

Berlin, James A. *Rhetoric and Reality: Writing Instruction in American Colleges, 1900–1985*. Carbondale: Southern Illinois UP, 1987.

Berry, Thomas. *The Dream of the Earth*. San Francisco: Sierra Club, 1988.

Berry, Wendell. *The Unsettling of America*. San Francisco: Sierra Club, 1977.

Bertin, Jacques. *Graphics and Graphic Information Processing*. Trans. William J. Berg and Paul Scott. New York: de Gruyter, 1981.

Bettelheim, Bruno. *The Uses of Enchantment: The Meaning and Importance of Fairy Tales*. New York: Knopf, 1976.

Bookchin, Murray. *The Ecology of Freedom: The Emergence and Dissolution of Hierarchy*. Palo Alto, CA: Chelsea, 1982.

——— [Lewis Herber, pseud.]. *Our Synthetic Environment*. New York: Knopf, 1962.

———. *Remaking Society: Pathways to a Green Future*. Boston: South End, 1990.

Booth, William. "Johnny Appleseed and the Greenhouse." *Science* 242 (1988): 19–20.

Bosso, Christopher J. *Pesticides and Politics: The Life Cycle of a Public Issue.* Pittsburgh: U of Pittsburgh P, 1987.

Bramwell, Anna. *Ecology in the 20th Century: A History.* New Haven: Yale UP, 1989.

Brand, David. "Is the Earth Warming Up?" *Time* 4 July 1988: 18.

Brennan, Andrew. *Thinking about Nature: An Investigation of Nature, Value and Ecology.* Athens: U of Georgia P, 1988.

Brookes, Warren T. "Stirring Up Environmentalist Hysteria." *Memphis Commercial-Appeal* 17 Sept. 1989: A4.

Brooks, Paul. *Speaking for Nature: How Literary Naturalists from Henry Thoreau to Rachel Carson Have Shaped America.* San Francisco: Sierra Club, 1980.

Brosnan, James W. "Velsicol Blamed for Tainted Foods." *Memphis Commercial-Appeal* 30 Aug. 1989: A13.

Brown, Lester. *Building a Sustainable Society.* New York: Norton, 1981.

———. *Seeds of Change: The Green Revolution and Development in the 1970's.* New York: Praeger, 1970.

Brown, Lester, et al. *State of the World 1988: A Worldwatch Institute Report on Progress toward a Sustainable Society.* New York: Norton, 1988.

———. *State of the World 1989: A Worldwatch Institute Report on Progress toward a Sustainable Society.* New York: Norton, 1989.

———. *State of the World 1990: A Worldwatch Institute Report on Progress toward a Sustainable Society.* New York: Norton, 1990.

Brown, Michael. "The Zeal of Disapproval." *Oceans* June 1987: 36–41.

Brown, Richard Harvey. *Society as Text.* Chicago: U of Chicago P, 1987.

Burke, Kenneth. *A Grammar of Motives.* Berkeley: U of California P, 1969.

———. *A Rhetoric of Motives.* Berkeley: U of California P, 1969.

Callenbach, Ernest. *Ecotopia: The Novel of Your Future.* 1975. New York: Bantam, 1982.

Callinicos, Alex. *Marxism and Philosophy.* New York: Oxford UP, 1985.

Carin, Arthur A., and Robert B. Sund. *Teaching Science Through Discovery.* 5th ed. Columbus, OH: Merrill, 1985.

Carson, Rachel. *Silent Spring.* New York: Fawcett Crest, 1962.

Charlier, Tom. "Pollution a Price as Town Prospers." *Memphis Commercial-Appeal* 18 June 1989: A1, 5.

Cheney, Patrick, and David Schleicher. "Redesigning Technical Reports: A Rhetorical Editing Method." *Journal of Technical Writing and Communication* 14 (1984): 317–37.

Church, George J. "Garbage, Garbage, Everywhere." *Time* 5 Sept. 1988: 81–82.

Clarke, David D. *Language and Action: A Structural Model of Behaviour.* Oxford: Pergamon, 1983.

Collins, Beverly, and S. T. A. Pickett. "Influence of Canopy Opening on the Environment and Herb Layer in a Northern Hardwoods Forest." *Vegetatio* 70 (1987): 3–10.
Commoner, Barry. *The Closing Circle: Nature, Man, and Technology*. New York: Knopf, 1971.
———. *Making Peace with the Planet*. New York: Pantheon, 1990.
———. *The Politics of Energy*. New York: Knopf, 1979.
———. *The Poverty of Power: Energy and the Economic Crisis*. New York: Knopf, 1976.
———. "A Reporter at Large: The Environment." *New Yorker* 15 June 1987: 46–71.
———. *Science and Survival*. New York: Viking, 1967.
Corbett, Edward P. J. *Classical Rhetoric for the Modern Student*. 2nd ed. New York: Oxford UP, 1971.
Cox, Harvey. *Religion and the Secular City*. New York: Simon, 1984.
Daly, Herman E. *Steady-State Economics: The Economics of Biophysical Equilibrium and Moral Growth*. San Francisco: Freeman, 1977.
Daly, Herman E., and John B. Cobb, Jr. *For the Common Good: Redirecting the Economy toward Community, the Environment, and a Sustainable Future*. Boston: Beacon, 1989.
Dawkins, Richard. *The Selfish Gene*. New York: Oxford UP, 1976.
Devall, Bill. *Simple in Means, Rich in Ends: Practicing Deep Ecology*. Salt Lake City: Peregrine Smith, 1988.
Devall, Bill, and George Sessions. *Deep Ecology: Living as if Nature Mattered*. Salt Lake City: Peregrine Smith, 1985.
DeVoto, Bernard. "The West Against Itself." *Voices for the Earth: A Treasury of the Sierra Club Bulletin*. Ed. Ann Gilliam. San Francisco: Sierra Club. 1979: 386–91.
Dewey, John. *Human Nature and Conduct*. New York: Modern Library, 1930.
———. *Liberalism and Social Action*. New York: Putnam's, 1935.
———. *The Quest for Certainty*. New York: Putnam's, 1929.
Dickson, David. "Bomb Scandal Highlights French Testing." *Science* 229 (1985): 948–49.
Dreyfus, Daniel A., and Helen M. Ingram. "The National Environmental Policy Act: A View of Intent and Practice." *Natural Resources Journal* 16 (1976): 244–62.
Dubos, René. "The Limits of Adaptability." *The Environmental Handbook*. Ed. Garrett De Bell. New York: Ballantine, 1970. 27–30.
———. *The Wooing of the Earth: New Perspectives on Man's Use of Nature*. New York: Scribner's, 1980.
Dykstra, Peter. "Greenpeace." *Environment* July–Aug. 1986: 5, 44–45.

Dyson, Freeman. *Disturbing the Universe.* New York: Harper, 1979.
Eagleton, Terry. *Literary Theory: An Introduction.* Minneapolis: U of Minnesota P, 1983.
EarthWorks Group. "Simple Ways You Can Help the Earth." *Reader's Digest* June 1990: 135–38.
Ehrlich, Paul R., and John P. Holdren. "Impact of Population Growth." *Toward a Steady-State Economy.* Ed. Herman Daly. San Francisco: Freeman, 1973. 76–89.
Ela, Jonathan. "From Sea to Shining Sea or through the Rockies at 31 Knots." *Voices for the Earth: A Treasury of the Sierra Club Bulletin.* Ed. Ann Gilliam. San Francisco: Sierra Club. 1979: 366–68.
Evernden, Neil. *The Natural Alien: Humankind and Environment.* Toronto: U of Toronto P, 1985.
Fairfax, Sally K. "A Disaster in the Environmental Movement." *Science* 199 (1978): 743–48.
Fisher, Walter R. *Human Communication as Narration: Toward a Philosophy of Reason, Value, and Action.* Columbia: U of South Carolina P, 1987.
Fleck, Ludwik. *Genesis and Development of a Scientific Fact.* Chicago: U of Chicago P, 1979.
"The Forecast: Hazy and Puzzling." *Time* 6 Feb. 1989: 57.
Foreman, Dave. "Earth First!" *The Progressive* Oct. 1981: 39–42.
———, ed. *Ecodefense: A Field Guide to Monkeywrenching.* Forward! [*sic*] by Edward Abbey. Tucson: Earth First!, 1985.
Fox, Stephen. *John Muir and His Legacy: The American Conservation Movement.* Boston: Little, 1981.
Friedman, Sharon M., Sharon Dunwoody, and Carol L. Rogers, eds. *Scientists and Journalists: Reporting Science as News.* New York: Free, 1986.
Friesema, H. Paul, and Paul J. Culhane. "Social Impacts, Politics, and the Environmental Impact Statement Process." *Natural Resources Journal* 16 (1976): 339–56.
Fuller, Steve. *Social Epistemology.* Bloomington: Indiana UP, 1988.
Garelik, Glenn. "A Breath of Fresh Air." *Time* 28 Sept. 1987: 35.
Georgescu-Roegen, Nicholas. "The Entropy Law and the Economic Problem." *Toward a Steady-State Economy.* Ed. Herman E. Daly. San Franciso: Freeman, 1973. 37–49.
Goldsmith, Maurice. *The Science Critic: A Critical Analysis of the Popular Presentation of Science.* London: Routledge, 1986.
Gorz, André. *Ecology as Politics.* Trans. Patsy Vigderman and Jonathan Cloud. Montreal: Black Rose, 1980.
Graber, Linda. *Wilderness as Sacred Space.* Washington, DC: Assn. of American Geologists, 1976.

Graham, Frank, Jr. *Since Silent Spring*. Boston: Houghton, 1970.
Gross, Alan G. *The Rhetoric of Science*. Cambridge: Harvard UP, 1990.
Gutzke, William H. N., et al. "Influence of the Hydric and Thermal Environments on Eggs and Hatchlings of Painted Turtles (Chrysemys Picta)." *Herpetologica* 43 (1987): 393–404.
Habermas, Jürgen. *Legitimation Crisis*. Trans. Thomas McCarthy. Boston: Beacon, 1975.
———. *The Philosophical Discourse of Modernity*. Trans. Frederick G. Lawrence. Cambridge: MIT Press, 1987.
———. *The Theory of Communicative Action*. Trans. Thomas McCarthy. 2 vols. Boston: Beacon, 1984–87.
Hacking, Ian. *Representing and Intervening: Introductory Topics in the Philosophy of Natural Science*. Cambridge: Cambridge UP, 1983.
Haines, Lionel. "The Green Barrage." *Business Month* Dec. 1989: 70–74.
Hammond, K. R., et al. "Fundamental Obstacles to the Use of Scientific Information in Public Policy Making." *Technological Forecasting and Social Change* 24 (1983): 287–97.
Hansen, James, et al. "Global Climate Changes as Forecast by Goddard Institute for Space Studies Three-Dimensional Model." *Journal of Geophysical Research* 93 (1988): 9341–64.
Hardin, Garrett. "Nobody Ever Dies of Overpopulation." *Voices for the Earth: A Treasury of the Sierra Club Bulletin*. Ed. Ann Gilliam. San Francisco: Sierra Club, 1979. 486–87.
———. "The Tragedy of the Commons." *The Environmental Handbook*. Ed. Garrett De Bell. New York: Ballantine, 1980. 31–50.
Harris, Ora Fred, Jr. "Communicating the Hazards of Toxic Substance Exposure." *Journal of Legal Education* 39 (1989): 97–112.
Harwood, Michael. "Daredevils for the Environment." *New York Times Magazine* 2 Oct. 1988. 72–76.
Hays, Samuel P. *Beauty, Health, and Permanence: Environmental Politics in the United States: 1955–1985*. Cambridge: Cambridge UP, 1987.
———. *Conservation and the Gospel of Efficiency: The Progressive Conservation Movement, 1890–1920*. Cambridge: Harvard UP, 1959.
Heidegger, Martin. *Being and Time*. Trans. John Macquarrie and Edward Robinson. New York: Harper, 1962.
———. "Building Dwelling Thinking." *Basic Writings*. Ed. D. F. Krell. New York: Harper, 1977. 320–39.
Herron, Matt. "Not Altogether Quixotic Face-Off with Soviet Whale Killers in the Pacific." *Smithsonian* Aug. 1976: 22–31.
Hood, Leslie, and James K. Morgan. "Whose Home on the Range?" *Voices for the Earth: A Treasury of the Sierra Club Bulletin*. Ed. Ann Gilliam. San Francisco: Sierra Club, 1979. 400–405.

Horkheimer, Max, and Theodor W. Adorno. *Dialectic of Enlightenment*. Trans. John Cumming. New York: Continuum, 1988.
Houp, Kenneth, and Thomas Pearsall. *Reporting Technical Information*. 5th ed. New York: Macmillan, 1984.
Ihde, Don. *Technology and the Lifeworld: From Garden to Earth*. Bloomington: Indiana UP, 1990.
Illich, Ivan. *Tools for Conviviality*. New York: Harper, 1973.
Kane, Joe. "Mother Nature's Army." *Esquire* Feb. 1987: 98–102.
Kaufman, Elizabeth. "Earth-Saving: Here is a Gang of Real Environmental Extremists." *Audubon* July 1982: 116–20.
Kerr, Richard A. "Is the Greenhouse Here?" *Science* 239 (1988): 559–61.
———. "Ozone Hole Bodes Ill for the Globe." *Science* 241 (1988): 785–86.
———. "Report Urges Greenhouse Action Now." *Science* 241 (1988): 23–24.
———. "Winds, Pollutants Drive Ozone Hole." *Science* 238 (1987): 156–58.
Killingsworth, M. Jimmie. "Can an English Teacher Contribute to the Energy Debate?" *College English* 43 (1981): 581–86.
———. "Guest Editorial: Thingishness and Objectivity in Technical Style." *Journal of Technical Writing and Communication* 17 (1987): 105–13.
———. "Rhetoric and Relevance in Technical Writing." *Journal of Technical Writing and Communication* 16 (1986): 287–96.
———. "Toward a Rhetoric of Technological Action." *International Technical Communication Conference Proceedings* 34 (1987): RET-136–37.
———. *Whitman's Poetry of the Body: Sexuality, Politics, and the Text*. Chapel Hill: U of North Carolina P, 1989.
Killingsworth, M. Jimmie, and Michael K. Gilbertson. "How Can Text and Graphics Be Integrated Effectively?" *Solving Problems in Technical Writing*. Ed. Lynn Beene and Peter White. New York: Oxford UP, 1988. 130–49.
Killingsworth, M. Jimmie, and Dean Steffens. "Effectiveness in the Environmental Impact Statement: A Study in Public Rhetoric." *Written Communication* 6 (1989): 155–80.
Kinneavy, James. *A Theory of Discourse*. New York: Norton, 1971.
Klaine, S. J., et al. "Characterization of Agricultural Nonpoint Pollution: Nutrient Loss and Erosion in a West Tennessee Watershed." *Environmental Toxicology and Chemistry* 7 (1988): 601–7.
———. "Characterization of Agricultural Nonpoint Pollution: Pesticide Migration in a West Tennessee Watershed." *Environmental Toxicology and Chemistry* 7 (1988): 609–14.
Koshland, Donald E., Jr. "Inexorable Laws and the Ecosystem." *Science* 237 (1987): 9.

Kuhn, Thomas S. *The Structure of Scientific Revolutions*. 2nd ed. Chicago: U of Chicago P, 1970.

Laclau, Ernesto, and Chantal Mouffe. *Hegemony and Socialist Strategy: Toward a Radical Democratic Politics*. London: Verso, 1985.

Lakatos, Irme. *The Methodology of Scientific Research Programmes*. Ed. John Worrall and Gregory Currie. Cambridge: Cambridge UP, 1978.

Lanham, Richard A. *Analyzing Prose*. New York: Scribner's, 1983.

Larrain, Jorge. *The Concept of Ideology*. Athens: U of Georgia P, 1979.

Latour, Bruno. *Science in Action: How to Follow Scientists and Engineers Through Society*. Cambridge: Harvard UP, 1987.

Lemonick, Michael D. "Feeling the Heat." *Time* 2 Jan. 1989: 36–41.

———. "Shrinking Shores." *Time* 10 Aug. 1987: 38–47.

Leopold, Aldo. *Sand County Almanac*. New York: Ballantine, 1966.

Linden, Eugene. "Big Chill for the Greenhouse." *Time* 31 Oct. 1988: 90–91.

———. "Putting the Heat on Japan." *Time* 10 July 1989: 50–52.

Lovelock, James. *The Ages of Gaia: A Biography of Our Living Earth*. New York: Bantam, 1990.

Luccitta, Ivo, David Schleicher, and Patrick Cheney. "Of Price and Prejudice: The Importance of Being Earnest about Environmental Impact Statements." *Geology* 9 (1981): 590–91.

Lyotard, Jean-François. *The Postmodern Condition: A Report on Knowledge*. Trans. Geoff Bennington and Brian Massumi. Foreword by Fredric Jameson. Minneapolis: U of Minnesota P, 1984.

MacIntyre, Alasdair. *Beyond Virtue*. 2nd ed. Notre Dame, IN: U of Notre Dame P, 1984.

McPhee, John. *Encounters with the Archdruid*. New York: Farrar, 1971.

Magnan, George A. "Industrial Illustrating and Layout." *Handbook of Technical Writing Practices*. Ed. Stello Jordon. 2 vols. New York: Wiley Interscience, 1971.

Malanowski, Jamie. "Monkey-Wrenching Around." *Nation* 2 May 1987: 568–70.

Manes, Christopher. *Green Rage: Radical Environmentalism and the Unmaking of Civilization*. Boston: Little, 1990.

Manicas, Peter T. *A History and Philosophy of the Social Sciences*. New York: Blackwell, 1987.

Marcuse, Herbert. *One-Dimensional Man: Studies in the Ideology of Advanced Industrial Society*. Boston: Beacon, 1964.

Marquardt, Sandra. *Exporting Banned Pesticides: Fueling the Circle of Poison*. Washington, DC: Greenpeace, 1989.

Marshall, Eliot. "EPA's Plan for Cooling the Global Greenhouse." *Science* 243 (1989): 1544–45.

Marth, Del. " 'I Must Search for Balance': In-Depth Interview with Secretary Watt." *Nation's Business* Sept. 1981: 36–41.

Martinez-Alier, Juan, and Klaus Schlupmann. *Ecological Economics: Energy, Environment, and Society*. Oxford: Blackwell, 1987.

Marx, Karl. *Economic and Philosophic Manuscripts of 1844*. London: Lawrence, 1959.

Marx, Karl, and Friedrich Engels. *The German Ideology*. Ed. R. Pascal. New York: International, 1947.

May, Clifford D. "Pollution Ills Stir Support for Environmental Groups." *New York Times* 21 Aug. 1988: 20Y.

Miller, Carolyn R. "Environmental Impact Statements and Some Modern Traditions of Communication." *International Technical Communication Conference Proceedings* 28 (1981): E-67–69.

———. "Genre as Social Action." *Quarterly Journal of Speech* 70 (1984): 151–167.

Miller, Robert L. "From the Publisher." *Time* 2 Jan. 1989: 3.

Monroe, Martha C., and Stephen Kaplan. "When Words Speak Louder Than Actions: Environmental Problem Solving in the Classroom." *Journal of Environmental Education* 19 (1987): 38–41.

Moore, Peter D. "Blow, Blow Thou Winter Wind." *Nature* 336 (1988): 313.

Morris, Charles. *Writings on the General Theory of Signs*. Ed. Thomas A. Sebeok. The Hague: Mouton, 1971.

Muir, John. *Our National Parks*. Boston: Houghton, 1901.

Mumford, Lewis. *The Myth of the Machine: The Pentagon of Power*. New York: Harcourt, 1970.

———. *The Myth of the Machine: Technics and Human Development*. New York: Harcourt, 1967.

Mumme, Ronald L., W. D. Koenig, and F. A. Pitelka. "Mate Guarding in the Acorn Woodpecker: Within-Group Reproductive Competition in a Cooperative Breeder." *Animal Behavior* 31 (1983): 1094–1106.

———. "Reproductive Competition in the Communal Acorn Woodpecker: Sisters Destroy Each Other's Eggs." *Nature* 306 (1983): 583–84.

Naess, Arne, and David Rothenberg. *Ecology, Community and Lifestyle: Outline of an Ecosophy*. Cambridge: Cambridge UP, 1989.

Nash, Roderick. *Wilderness and the American Mind*. 3rd ed. New Haven: Yale UP, 1982.

Nietzsche, Friedrich. "On Truth and Falsity in Their Ultramoral Sense (1873)." *The Complete Works of Friedrich Nietzsche*. Ed. Oscar Levy. Vol. 2. New York: Russell, 1964. 171–92.

Noble, David F. *America by Design: Science, Technology, and the Rise of Corporate Capitalism*. Oxford: Oxford UP, 1977.

Norstog, Knut, and Andrew J. Meyerriecks. *Biology*. Columbus, OH: Merrill, 1983.

Ong, Walter J. "Voice as a Summons for Belief." *The Barbarian Within*. New York: Macmillan, 1962. 49–67.

Ortega y Gasset, José. *Historical Reason*. Trans. Philip W. Silver. New York: Norton, 1984.

Orwell, George. *1984*. New York: NAL, 1981.

———. "Politics and the English Language." *The Norton Reader*. Ed. Arthur M. Eastman, et al. 5th ed. New York: Norton, 1980. 369–80.

Otto, James H., and Albert Towle. *Modern Biology*. New York: Holt, 1985.

Packard, Vance. *The Hidden Persuaders*. New York: McKay, 1957.

Pain, Stephanie. "No Escape from the Global Greenhouse." *New Scientist* 12 Nov. 1988: 38–43.

Palmer, Parker J. "Community, Conflict, and Ways of Knowing." *Change* Sept.–Oct. 1987: 20–25.

Parfit, Michael. "Earth First!ers Wield a Mean Monkey Wrench." *Smithsonian* Apr. 1990: 184–204.

Parsons, Howard L. *Marx and Engels on Ecology*. Westport, CT: Greenwood, 1977.

Perlman, Eric. "Confrontation: Greenpeace Foundation Puts Itself on the Line." *Oceans* July–Aug. 1977: 58–61.

Petersen, David. "The Plowboy Interview: Dave Foreman: No Compromise in Defense of Mother Earth." *Mother Earth News* Jan.–Feb. 1985: 16–22.

Pike, Kenneth. *Linguistic Concepts: An Introduction to Tagmemics*. Lincoln: U of Nebraska P, 1982.

P. M. *Bolo'bolo*. New York: Semiotext(e), 1985.

Popper, Karl R. *The Logic of Scientific Discovery*. London: Hutchinson, 1959.

———. *The Open Society and Its Enemies*. 2 vols. Princeton: Princeton UP, 1966.

Porritt, Jonathon. *Seeing Green: The Politics of Ecology Explained*. New York: Blackwell, 1985.

Poster, Mark. *Critical Theory and Poststructuralism: In Search of a Context*. Ithaca, NY: Cornell UP, 1989.

Potter, Van Rensselaer. *Global Bioethics: Building on the Leopold Legacy*. East Lansing: Michigan State UP, 1988.

Prelli, Lawrence J. A *Rhetoric of Science: Inventing Scientific Discourse*. Columbia: U of South Carolina P, 1989.

Prescott, W. R. "Governance and Global Warming." *In Context* (Summer 1989): 38–43.

Price, Derek J. de Solla. *Little Science, Big Science . . . and Beyond*. New York: Columbia UP, 1986.

Purl, Mara. "Greenpeace Pirates Save the Whales." *Rolling Stone* 13 July 1978: 8, 24–33.

Putnam, Hilary. *The Many Faces of Realism*. LaSalle, IL: Open Court, 1987.

Ramanathan, V. "The Greenhouse Theory of Climate Change: A Test by an Inadvertent Global Experiment." *Science* 240 (1988): 293–99.

Reiger, George. "April Foolishness." *Field and Stream* Apr. 1986: 14–16.

ReVelle, Penelope, and Charles ReVelle. *The Environment: Issues and Choices for Society*. 3rd ed. Boston: Jones, 1988.

Reynolds, John C., and Lutian R. Wootton. "Educational Trends: Impact on Schooling." *Contemporary Education* 60 (1988): 6–11.

Rorty, Richard. *Contingency, Irony, and Solidarity*. Cambridge: Cambridge UP, 1989.

———. *Philosophy and the Mirror of Nature*. Princeton: Princeton UP, 1979.

Rosenblatt, Roger. "Why Reagan Is Funny and Watt Not." *Time* 17 Oct. 1983: 100.

Rossini, Frederick A., and Alan L. Porter, "Public Participation and Professionalism in Impact Assessment." *Journal of Voluntary Action Research* 11.1 (1982): 24–33.

Roth, Charles E. "The Endangered Phoenix—Lessons from the Firepit." *Journal of Environmental Education* 19 (1988): 3–9.

Russell, Christine. "The View from the National Beat." *Scientists and Journalists: Reporting Science as News*. Ed. Sharon M. Friedman, Sharon Dunwoody, and Carol L. Rogers. New York: Free, 1986. 81–94.

Schell, Jonathan. *The Fate of the Earth*. New York: Knopf, 1982.

Schindler, David W. "Detecting Ecosystem Responses to Anthropogenic Stress." *Canadian Journal of Fisheries and Aquatic Science* 44, Suppl. 1 (1987): 6–25.

———. "The Impact Statement Boondoggle." *Science* 192 (1976): 509.

Schneider, Stephen H. "Both Sides of the Fence: The Scientist as Source and Author." *Scientists and Journalists: Reporting Science as News*. Ed. Sharon M. Friedman, Sharon Dunwoody, and Carol L. Rogers. New York: Free, 1986. 215–22.

———. *The Coevolution of Climate and Life*. San Francisco: Sierra Club, 1984.

———. *Global Warming: Are We Entering the Greenhouse Century?* San Francisco: Sierra Club, 1989.

———. "The Greenhouse Effect: Science and Policy." *Science* 243 (1989): 771–81.

Schumacher, E. F. *Small Is Beautiful: A Study of Economics as if People Mattered*. London: Blond, 1973.

"Secretary Watt Fires Back at His Critics." *Business Week* 24 Jan. 1983: 86.

Semlitsch, Raymond, and J. W. Gibbons. "Fish Predation in Size-Structured Populations of Treefrog Tadpoles." *Oecologia* [Berlin] 75 (1988): 321–26.

Sidney, Hugh. "The Big Dry." *Time* 4 July 1988: 12–15.

Simpson, J. C. "The Captain Who Caused a Furor." *Time* 2 Sept. 1985: 27.

Strain, Joy G., and Ronald L. Mumme. "Effects of Food Supplementation, Song Playback, and Temperature on Vocal Territorial Behavior of Carolina Wrens." *Auk* 105 (1988): 11–16.

Suro, Robert. "Polluters Are Fought Near Home." (*New York Times* News Service). *Memphis Commercial-Appeal* 2 July 1989: A1.

Teter, Harold, et al. *Living Things: An Introduction to Biology*. Teacher's edition. New York: Holt, 1985.

Toufexis, Anastasia. "The Dirty Seas." *Time* 1 Aug. 1988: 44–50.

Toulmin, Stephen. *The Return to Cosmology: Postmodern Science and the Theology of Nature*. Berkeley: U of California P, 1982.

Tufte, Edward R. *The Visual Display of Quantitative Information*. Cheshire, CT: Graphics, 1983.

Turnbull, Arthur T., and R. Baird. *The Graphics of Communication*. 3rd ed. New York: Holt, 1975.

United States Dept. of Agriculture and Dept. of the Interior. Forest Service. [U.S. Forest Service]. *Environmental Impact Statement for the Cibola National Forest Plan*. Albuquerque, NM: np, 1985.

United States Dept. of the Interior. Bureau of Land Management [U.S. BLM]. *Draft Environmental Impact Statement for the Proposed West Socorro Rangeland Management Program*. Socorro, NM: BLM, 1982.

———. *East Socorro Grazing Environmental Statement*. Socorro, NM: BLM, 1979.

———. *Editorial Management Handbook for Environmental Impact Statement and Environmental Assessment*. Washington, DC: Dept. of the Interior, 1980.

———. *New Mexico Statewide Wilderness Study: Draft EIS*. Albuquerque, NM: BLM, 1985.

Velsicol Chemical Corporation [Memphis, TN]. "Velsicol Calls Greenpeace Report 'Inaccurate.' " Company news release 29 Aug. 1989.

Weber, Max. *The Protestant Ethic and the Spirit of Capitalism*. Trans. Talcott Parsons. New York: Scribner's, 1958.

Wein, Gary, Stephen Kroeger, and Gary J. Pierce. "Lacustrine Vegetation Establishment within a Cooling Reservoir." Unpublished paper, 1987.

Wein, Gary, and William D. McCort. "Sources of Complexity in Wetland Migration." *Increasing Our Wetland Resources.* Ed. John Zelazny and J. Scott Feierabend. Washington, DC: National Wildlife Federation, 1988. 41–50.

Weiner, Jonathan. *The Next One Hundred Years: Shaping the Fate of Our Living Earth.* New York: Bantam, 1990.

"Whale of a War off Iceland: A Battle at Sea on the Eve of a Summit to Save the Leviathans." *Time* 9 July 1979: 45–47.

White, Hayden. "The Value of Narrativity in the Representation of Reality." *Critical Inquiry* 7 (1980): 5–27.

Williams, Joseph. *Style: Ten Lessons in Clarity and Grace.* 2nd ed. Glenview, IL: Scott, 1985.

World Commission on Environment and Development. *Our Common Future.* New York: Oxford UP, 1987.

Index

M. Jimmie Killingsworth is associate professor of English and director of writing programs at Texas A&M University. He is the author of *Whitman's Poetry of the Body: Sexuality, Politics, and the Text* (Chapel Hill: University of North Carolina Press, 1989) and numerous articles on American literature, rhetoric, and scientific and technical communication.

Jacqueline S. Palmer is a researcher and science teacher. The author of several articles on science education, she works as a research associate in the Department of Curriculum and Instruction, Texas A&M University.

Killingsworth and Palmer, who frequently collaborate in their research and writing on the cultural and educational effects of environmentalism, are married and have a daughter, Miki.